Springer Theses

Recognizing Outstanding Ph.D. Research

Aims and Scope

The series "Springer Theses" brings together a selection of the very best Ph.D. theses from around the world and across the physical sciences. Each volume has been selected for its scientific excellence by an internationally renowned institute or department. For greater accessibility to non-specialists, the published versions include an extended introduction, as well as a foreword by the student's supervisor explaining the special relevance of the work for the field. The series as a whole provides a valuable resource both for newcomers to the research fields, and for other scientists seeking detailed background information on special questions. Finally, it serves as an accredited documentation of the valuable contributions made by today's younger generation of scientists.

How to Apply

Theses prepared at internationally leading universities and institutes may be nominated for publication by supervisors with the endorsement of the head of department. Each such department is entitled to nominate one thesis per year. Enquiries and requests for a nomination form should be sent, together with the thesis, to Ms. Nathalie Berg (see link on right).

Eligibility Criteria

To qualify for consideration, theses should fulfil the following criteria:

- They must be written in good English.
- The topic should fall within the confines of Chemistry, Physics, Earth Sciences, Engineering and related interdisciplinary fields such as Materials, Nanoscience, Chemical Engineering, Complex Systems and Biophysics.
- The work reported in the thesis must represent a significant scientific advance.
- A maximum of 30% of the text may be a verbatim reproduction from the author's previous publications.
- They must have been examined and passed during the 12 months prior to nomination.
- The theses should have a clearly defined structure including an introduction accessible to new PhD students and scientists not expert in the relevant field.
- After acceptance, theses should be supplemented by a "Supervisor's Foreword" outlining the wider scientific relevance of the work.

Indexed by zbMATH.

Yuma Nakamura

Generation of Highly Controllable Ytterbium Atom Array

A Hybrid System of Nuclear Spin and Optical Clock Qubits Towards Quantum Error Correction

Doctoral Thesis accepted by
Kyoto University, Kyoto, Japan

Springer

Author
Dr. Yuma Nakamura
Department of Physics, Graduate School of Science
Kyoto University
Kyoto, Japan

Supervisor
Dr. Yoshiro Takahashi
Department of Physics, Graduate School of Science
Kyoto University
Kyoto, Japan

ISSN 2190-5053 ISSN 2190-5061 (electronic)
Springer Theses
ISBN 978-981-95-2835-6 ISBN 978-981-95-2836-3 (eBook)
https://doi.org/10.1007/978-981-95-2836-3

This Springer imprint is published by the registered company Springer Nature Singapore Pte Ltd.
The registered company address is: 152 Beach Road, #21-01/04 Gateway East, Singapore 189721, Singapore

Supervisor's Foreword

Recently, many efforts have been made to realize quantum computers, the next-generation computers based on the principle of quantum mechanics. In the last decade, neutral atoms in a reconfigurable optical tweezer array have emerged as a promising platform for quantum computing, offering high scalability and connectivity of the qubits, which are preferable properties in quantum error correction (QEC) schemes. In neutral-atom-based quantum computing architecture, one of the difficulties in implementing QEC arises from the readout of auxiliary qubits, which are called ancilla qubits and are used for error detection, without losing the ancilla qubits and disturbing the data qubits state and coherence. Since the auxiliary degrees of freedom of the system play a central role in most QEC protocols, recent studies on neutral atom arrays focus on the efficient preparation, control, and detection of ancilla qubits to achieve QEC.

To address this challenge, in the present thesis, Dr. Yuma Nakamura's work demonstrates a hybrid atom array system of dual-isotope alkali-earth-like ytterbium (Yb) atoms. While previous research using dual-species and dual-isotope alkali atom array has demonstrated the basic advantages of this type of experimental platforms, Dr. Yuma Nakamura's new architecture is different in that it consists of two different kinds of atomic qubits, namely, fermionic isotope ^{171}Yb offering nuclear spin qubits, and bosonic isotope ^{174}Yb providing optical clock qubits, where the former acts as data qubits with second-order coherence and the latter as ancilla qubits capable of nondestructive qubit readout. The efficient preparation of a dual-Yb single-atom array is demonstrated by using a photoassociation process—converting two atoms to a molecule—to remove the simultaneous trapping of both isotopes. Importantly, Dr. Yuma Nakamura's work evaluates the crosstalk between the qubits by assessing the influence on the nuclear spin qubit coherence from the optical clock qubits imaging, observing that the coherence is not degraded even under irradiation of imaging light for ^{174}Yb, highlighting the potential of this system for the implementation of QEC. This work opens new avenues for developing a hybridized qubit system based on neutral atom architecture, facilitating fault-tolerant quantum computing.

In addition, in neutral-atom-based quantum computing architecture, the implementation of two-qubit gates, crucial for universal quantum computation, is achieved through interactions between Rydberg atoms, where the atom's valence electron is excited to an orbital with a large principal quantum number. Ytterbium atom presents a promising avenue for realizing high-fidelity two-qubit gates via single-photon Rydberg excitation from a metastable excited state. This approach addresses the limitations, such as the scattering from the intermediate state during Rydberg excitation, encountered with alkali atoms in earlier studies. In order to realize such Rydberg excitation, the development of a high-power ultraviolet (UV) laser is required.

Dr. Yuma Nakamura's work reports the development of a high-power UV laser at a wavelength of 325 nm for Rydberg excitation of single Yb atoms trapped in optical tweezer arrays. The laser system consists of a 1300 nm seed laser, a Raman fiber amplifier, and two-stage second harmonic generation to convert the wavelength to 325 nm, achieving a maximum output power exceeding 800 mW at 325 nm. This capability has the potential to enable 10 MHz-order fast Rydberg excitation, addressing Rydberg state decay, a primary limitation in the fidelity of two-qubit gates.

For high-fidelity quantum operations, it is crucial to manage laser phase noise. To address this, Dr. Yuma Nakamura's work has incorporated a high-finesse Fabry-Pérot cavity between the seed laser and the Raman fiber amplifier, achieving approximately 20 dB suppression of phase noise in the MHz range, anticipating operation fidelities surpassing the threshold required for quantum error correction. Using the developed UV laser, Dr. Yuma Nakamura's work demonstrates coherent excitation from the metastable $(6s6p)^3\mathrm{P}_2$ state to the $(6s71s)^3\mathrm{S}_1$ Rydberg state of $^{174}\mathrm{Yb}$, and in particular, successfully observes Rabi oscillations between these states with a Rabi frequency of $2\pi \times 2.13$ MHz, marking an essential first step towards two-qubit operations in Yb atom arrays. This work establishes a solid foundation for the realization of high-fidelity, Rydberg-mediated two-qubit gates, advancing the prospects for fault-tolerant quantum computation.

Dr. Yuma Nakamura's work was nominated as an outstanding Ph.D. Thesis for publication in the Springer Theses series by the Department of Physics, Graduate School of Science, Kyoto University.

Kyoto, Japan
June 2025

Yoshiro Takahashi

Preface

Neutral atoms trapped in optical tweezer arrays have emerged as a prominent platform for quantum simulation and computation. Their attractive features, such as scalable qubit arrays and flexible qubit connectivity enabled by moving tweezers, make them particularly suitable for implementing quantum error correction (QEC) toward fault-tolerant quantum computation. However, performing mid-circuit measurements—reading out ancilla qubits without disturbing data qubits—remains challenging. This difficulty arises from readout methods requiring global probe beam irradiation, which can induce heating and decoherence in the atom array.

In this study, we explore the potential of a dual-isotope ytterbium (Yb) atom array for mid-circuit measurements. The dual-Yb system consists of fermionic ^{171}Yb and bosonic ^{174}Yb isotopes. The nuclear spin qubit encoded in the ground state of ^{171}Yb serves as a data qubit, offering second-order long coherence times and MHz-scale fast qubit control. Meanwhile, the optical clock qubit in ^{174}Yb allows non-destructive readout, making it a promising candidate for the ancilla qubit.

We first achieved a dual-Yb system in a 10×10 optical tweezer array, obtaining a loading probability of 21(6)% for both isotopes while reducing dual occupancy, where both isotopes occupy the same site, to 0.2(7)% by optimizing the frequency of the light-assisted collision beam. Starting from this randomly loaded array, we constructed defect-free dual-Yb arrays using tweezer rearrangement, forming a robust starting point for quantum computation.

Next, we quantified the effect of imaging light for ^{174}Yb on the coherence of nuclear spin qubits in ^{171}Yb. By comparing Hahn-echo signals with and without imaging, we found that for a 20 ms exposure to imaging light, nuclear spin coherence was preserved at 99.1(1.8)% relative to the case without imaging. Simultaneously, we achieved a 174Yb detection fidelity of 0.9992 and a survival probability after imaging of 0.988.

In addition, we developed a high-power ultraviolet laser system at a wavelength of 325 nm and output power of 800 mW. Using this laser, we observed Rabi oscillations between the metastable $(6s6p)^3P_2$ and the Rydberg state $(6s71s)^3S_1$, achieving a Rabi frequency of $2\pi \times 2.13(3)$ MHz. This result is a foundational step toward implementing high-fidelity two-qubit gates mediated by Rydberg interactions.

By integrating the dual-Yb atom array's non-destructive readout capabilities with fast Rydberg excitation for two-qubit gates, our findings represent significant progress toward QEC and fault-tolerant quantum computing.

Kyoto, Japan Yuma Nakamura

Parts of this thesis have been published in the following journal articles

- Y. Nakamura, T. Kusano, R. Yokoyama, K. Saito, K. Higashi, N. Ozawa, T. Takano, Y. Takasu, and Y. Takahashi, Hybrid atom tweezer array of nuclear spin and optical clock qubits, Physical Review X **14**, 041062 (2024).
- Y. Nakamura, N. Ozawa, T. Kusano, R. Yokoyama, K. Shibata, T. Takano, Y. Takasu, Y. Takahashi, Development of a high-power ultraviolet laser system and observation of fast coherent Rydberg excitation of ytterbium, Journal of Physical Society of Japan **94**, 014301 (2025).

Acknowledgements This dissertation would not have been possible without the guidance and support of many individuals, especially the members of the Quantum Optics Laboratory at Kyoto University. I would like to take this opportunity to express my deepest gratitude to all those who have helped me throughout this journey.

First and foremost, I am deeply grateful to my academic advisor, Prof. Yoshiro Takahashi, who has guided me over the past six years since my fourth year as an undergraduate. Under his mentorship, I have acquired many essential skills as a researcher, including how to conduct research, write logically and persuasively, give engaging and clear presentations, and cultivate the right attitude toward scientific inquiry. I am especially thankful for the opportunity to work on the fascinating topic of neutral atom quantum computing in such a cutting-edge research environment as the Takahashi Laboratory at Kyoto University.

I would like to extend my sincere appreciation to Assoc. Prof. Yosuke Takasu for his tremendous support in building experimental apparatus, designing electrical circuits, and mastering experimental techniques. His practical advice, grounded in extensive experience, taught me many important aspects of experimentation that are not written in textbooks or papers.

I am also grateful to Assoc. Prof. Tetsushi Takano for his invaluable guidance, particularly in laser development. I benefited greatly from his insights during the development of the high-power 325 nm laser. He generously shared his knowledge of lasers and electronics, which was instrumental to the progress of my research.

I wish to thank the lab secretaries, Kumiko Terakawa and Miki Yasuhara, for their indispensable administrative support. They consistently handled various essential procedures—such as equipment orders and travel applications—with promptness and care. Even when faced with paperwork issues or complicated procedures, they always offered warm and reliable assistance.

I would also like to thank the members of the Tweezer group for their constant support in many aspects of my research. Associate Professor Kosuke Shibata provided numerous suggestions regarding the improvement of the experimental environment and the direction of experiments. I am especially grateful for the quantitative discussions we had, supported by numerical simulations. Associate Professor Toshihiko Shimazaki shared many valuable insights based on his international experience. I always appreciated his engaging conversations and the welcoming atmosphere he created, which allowed me to consult him whenever I faced difficulties. Naoya Ozawa played a vital role in managing many essential tasks within the group, and I believe our research would not have progressed as smoothly without his contributions—especially in the development of the 325 nm laser. Toshi Kusano has been my closest collaborator. I have truly enjoyed our discussions and admired his thorough and thoughtful approach to research. I hope we can continue working together and producing meaningful results. Rei Yokoyama, Keito Saito, and Koichiro Azuma assisted me with the development of the experimental setup and experiments.

I would also like to express my sincere thanks to former members of the Tweezer group—Nobuyuki Takei, Hideki Konishi, and Daichi Okuno—who guided me during my undergraduate and master's studies and taught me the fundamentals of experimentation. I especially appreciate Okuno, with whom I worked closely for about three years to build the foundation of this research. I am pleased that we continue to exchange ideas on research even today.

I am grateful to many other members of the Quantum Optics Laboratory. Assistant Professor Koki Ono provided me with detailed guidance on coherent control of nuclear spins, including how to calculate Rabi frequencies and light shifts, and how to choose optimal detunings. He also occasionally checked in on my work and offered valuable advice. I benefited greatly from discussions with Kantaro Honda and Taiki Ishiyama. They listened to my concerns and made lab life more enjoyable through their camaraderie and light-hearted conversations. Their continuous research achievements were a strong source of motivation for me.

I also received much guidance from researchers outside the laboratory. In particular, I would like to thank Sylvain de Léséleuc and Takafumi Tomita of the Institute for Molecular Science for their instruction on optical tweezer arrays and laser technology. I am especially grateful for the regular discussions we had, which provided many insights that shaped my research.

Lastly, I would like to express my heartfelt appreciation to my parents, who supported me throughout my long doctoral journey.

Competing Interests The author has no competing interests to declare that are relevant to the content of this manuscript.

Contents

Chapter 1
Introduction

1.1 Neutral Atoms in Optical Tweezers

Laser-cooled neutral atoms—cold atoms—[1–4] have been offering a versatile platform for quantum science and technology, owing to their high controllability and the cleanliness. Cold atoms in optical lattices have achieved significant breakthroughs, especially in quantum simulation of quantum many-body systems [5–9] and in quantum metrology [10–12], among many other fields.

In contrast to these top-down approaches, where the macroscopic behavior of the system is of primary importance, the bottom-up approach of quantum systems with individual atoms has also garnered significant interest [13]. Since the first demonstration of single-atom trapping in optical tweezers [14, 15] in 2001, the field has made rapid strides in the manipulation of individual atoms. These advancements include the cooling of single atoms to their motional ground state [16, 17], the observation of quantum phenomena [18, 19], and the implementation of quantum gate operations [20–27]. More recently, the realization of defect-free atom arrays using moving optical tweezers generated by acousto-optic deflectors [28, 29] or dynamical holographic tweezers [30] has unlocked significant potential for the optical tweezer array platform in quantum science.

The first application of neutral atoms in optical tweezer arrays is the quantum simulation of various quantum many-body systems [31]. By leveraging the long-range Rydberg interactions between atoms, it is possible to simulate spin models [32–41], topological models [42, 43], Hubbard models [44, 45], and other systems [46–50].

The second application is quantum metrology. The high controllability of individual atoms in optical tweezers enables the realization of high-precision atomic clocks [51–53]. Multibody entanglement generation [54] and spin squeezing [55, 56] using Rydberg interactions have been demonstrated, bringing quantum-enhanced metrology closer to realization.

The last, and arguably most prominent, application is quantum computing [57–61]. Neutral atoms, with inherently identical qubits encoded in atomic two-level systems,

Y. Nakamura, *Generation of Highly Controllable Ytterbium Atom Array*,
Springer Theses, https://doi.org/10.1007/978-981-95-2836-3_1

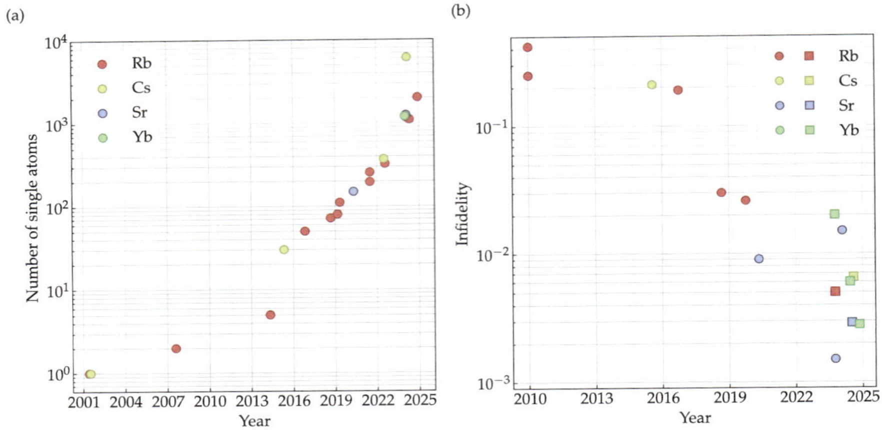

Fig. 1.1 Progress in neutral atom tweezer arrays. **a** The number of single atoms in an optical tweezer array. Data are taken from Refs. [14, 15, 21, 26, 28, 29, 36, 37, 52, 62–64, 66–71, 95–97]. **b** Gate fidelity in neutral atom-based quantum information processor. Circles and squares represent the Bell state fidelity and two-qubit gate fidelity, respectively. Data are taken from Refs. [22, 24, 25, 27, 54, 72–81]

offer long coherence times and fast qubit control, making them strong candidates for quantum computing. The number of qubits in optical tweezer arrays has been rapidly increasing [62–71] (Fig. 1.1a). Simultaneously, gate operation fidelities have improved significantly [54, 72–81] (Fig. 1.1b), making the platform competitive with alternatives like trapped ions [82, 83] and superconducting qubits [84–87]. Moreover, coherent atom transport via moving optical tweezers [88, 89] enables flexible, all-to-all qubit connectivity, facilitating the implementation of various quantum error correction codes [90–92]. To date, logical qubit encoding and quantum error detection have been successfully demonstrated [89, 93, 94].

1.2 Advantages of Alkali-Earth-Like Atoms

While alkali atoms, such as lithium [98, 99], sodium [100–102], potassium [103, 104], rubidium [14], and cesium [15], have been the primary choice for neutral atom experiments, alkali-earth(-like) atoms (AEA), including strontium [65, 105–110] and ytterbium [111–115], have recently emerged as attractive alternatives for optical tweezer experiments. The rich internal states of these atoms offer significant advantages for quantum computing.

The nuclear spin states of fermionic isotopes provide ideal qubits. Due to the absence of hyperfine coupling between the nuclear spin and electron–since the electronic angular momentum is canceled by two valence electrons–nuclear spin qubits exhibit coherence times that are three orders of magnitude longer than those of alkali atoms [107, 113, 116]. In particular, ytterbium-171 (^{171}Yb), with its nuclear spin of 1/2, features the simplest nuclear spin structure, enabling long coherence times while simultaneously allowing fast qubit control by utilizing the optical transition with the excited state $^3P_1$ with a hyperfine coupling [113].

Metastable excited states, such as the $^3P_0$ and $^3P_2$ states, offer long-lived optical clock qubits [117] and single-photon Rydberg excitation [74]. These features enable high-fidelity two-qubit gates [54, 75, 78–81]. The optical-metastable-ground (OMG) architecture, originally proposed for trapped ion systems [118], has been demonstrated with ytterbium atoms [75, 119], highlighting its potential for quantum error correction.

In the Rydberg states of AEAs, the inner valence electron is optically active, enabling Rydberg atom trapping [120] and light-shift-based Rydberg excitation control [121, 122]. Furthermore, fast Rydberg state detection via autoionization through excitation of the inner valence electron [74] facilitates erasure conversion [123], thereby enhancing the practicality of quantum error correction.

1.3 This Work

This work focuses on advancing quantum computing with neutral atoms, a promising platform for scalable, fault-tolerant quantum computing. A key challenge in this field is the implementation of quantum error correction, which requires high-fidelity qubit operations, robust qubit initialization, and mid-circuit measurements to detect and correct errors without disturbing the quantum state. To address these challenges, we develop a hybrid atom tweezer array based on dual-isotope ytterbium atoms, leveraging the unique properties of both fermionic and bosonic isotopes to create an integrated system of nuclear spin and optical clock qubits (Fig. 1.2).

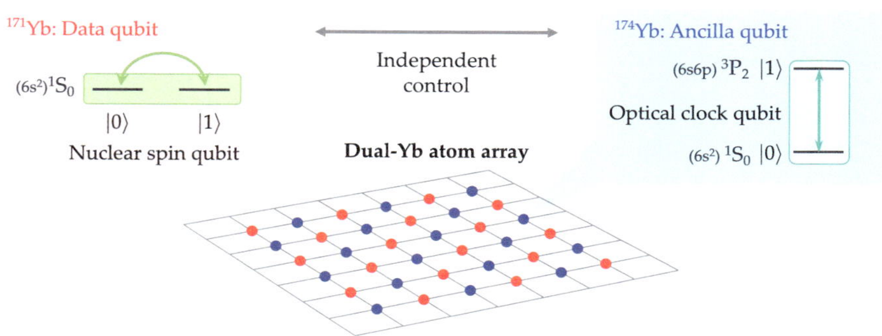

Fig. 1.2 Concept of the dual-Yb atom array. The fermionic isotope ^{171}Yb provides robust nuclear spin qubits for data qubits, while the bosonic isotope ^{174}Yb offers optical clock qubits for ancilla qubits. The hybrid atom tweezer array enables low crosstalk measurements of the ancilla qubits, facilitating mid-circuit measurements for quantum error correction

The fermionic isotope ^{171}Yb provides a robust nuclear spin qubit that benefits from long coherence times and fast control. These properties make ^{171}Yb an ideal candidate for data qubits that encode quantum information. In parallel, the bosonic isotope ^{174}Yb offers an optical clock qubit, which is advantageous for ancilla qubits due to its capacity for non-destructive, state-selective readout.

In the first stage of this work, we achieve efficient initialization and coherent control of nuclear spin qubits in ^{171}Yb atoms and optical clock qubits in ^{174}Yb atoms. We demonstrate coherent control over these qubits, such as state preparation and single-qubit rotations.

Building on these capabilities, we construct a hybrid atom tweezer array composed of ^{171}Yb and ^{174}Yb atoms. This system demonstrates tunable dual occupancy of the tweezer traps, indicating a possibility of the simultaneous manipulation of data and ancilla qubits within the same experimental setup. A significant milestone is achieved by demonstrating low-crosstalk measurements of ^{174}Yb ancilla qubits while preserving the quantum state of ^{171}Yb data qubits. This represents a key step toward implementing mid-circuit measurements, a critical component for quantum error correction that allows error detection and correction in real time without collapsing the quantum state of neighboring qubits.

In parallel with the development of the hybrid tweezer array, we design and construct a high-power ultraviolet (UV) laser system tailored for fast and high-fidelity Rydberg excitation of ytterbium atoms. The UV laser is stabilized using a filtering cavity to suppress phase noise, which is essential for achieving high-fidelity qubit operations via the Rydberg interaction. Using this system, we demonstrate fast and coherent Rydberg excitation as well as a spectroscopic signature of Rydberg interactions. These results establish the foundation for implementing Rydberg-mediated two-qubit gates, which are essential for scalable quantum computing architectures.

1.4 Outline

This thesis is organized as follows:

- In Chap. 2, we review the theoretical background of atom-light interactions, which play a crucial role in the coherent manipulation of nuclear spin and optical clock qubits, as well as in the trapping of ytterbium atoms.
- In Chap. 3, we describe the basic properties of ytterbium atoms, including their energy levels and transition characteristics. We also summarize the properties of Rydberg atoms, which enable long-range interactions between atoms.
- In Chap. 4, we present the experimental setup used in this work. This includes a description of experimental techniques such as optical tweezer generation, single-atom preparation, and detection.
- In Chap. 5, we demonstrate the coherent manipulation of nuclear spin and optical clock qubits in ytterbium atoms. Specifically, we show efficient initialization and coherent control of nuclear spin qubits in a ^{171}Yb atom array, as well as narrow-line

spectroscopy and coherent manipulation of optical clock qubits in a ^{174}Yb atom array.
- In Chap. 6, we report on the construction of a hybrid atom tweezer array composed of ^{171}Yb and ^{174}Yb atoms. Experimental results on crosstalk measurements between the isotopes are also presented.
- In Chap. 7, we describe the development of a high-power UV laser system tailored for the Rydberg excitation of ytterbium atoms and the observation of fast, coherent Rydberg excitation.
- In Chap. 8, we conclude this thesis and provide an outlook on future research directions.

References

1. D.J. Wineland, W.M. Itano, Phys. Rev. A Gen. Phys. **20**(4), 1521 (1979). https://doi.org/10.1103/physreva.20.1521. https://link.aps.org/doi/10.1103/PhysRevA.20.1521
2. F. Dalfovo, S. Giorgini, L.P. Pitaevskii, S. Stringari, Rev. Mod. Phys. **71**(3), 463 (1999). https://doi.org/10.1103/revmodphys.71.463. https://link.aps.org/doi/10.1103/RevModPhys.71.463
3. S. Giorgini, L.P. Pitaevskii, S. Stringari, Rev. Mod. Phys. **80**(4), 1215 (2008). https://doi.org/10.1103/RevModPhys.80.1215. https://link.aps.org/doi/10.1103/RevModPhys.80.1215
4. F. Schreck, K. van Druten, Nat. Phys. **17**(12), 1296 (2021). https://doi.org/10.1038/s41567-021-01379-w. https://www.nature.com/articles/s41567-021-01379-w
5. D. Jaksch, P. Zoller, Ann. Phys. **315**(1), 52 (2005). https://doi.org/10.1016/j.aop.2004.09.010. https://linkinghub.elsevier.com/retrieve/pii/S0003491604001782
6. I. Bloch, Nat. Phys. **1**(1), 23 (2005). https://doi.org/10.1038/nphys138. http://dx.doi.org/10.1038/nphys138
7. I. Bloch, J. Dalibard, W. Zwerger, Rev. Mod. Phys. **80**(3), 885 (2008). https://doi.org/10.1103/RevModPhys.80.885. https://link.aps.org/doi/10.1103/RevModPhys.80.885
8. C. Gross, I. Bloch, Science **357**(6355), 995 (2017). https://doi.org/10.1126/science.aal3837. http://dx.doi.org/10.1126/science.aal3837
9. F. Schäfer, T. Fukuhara, S. Sugawa, Y. Takasu, Y. Takahashi, Nat. Rev. Phys. **2**(8), 411 (2020). https://doi.org/10.1038/s42254-020-0195-3. http://www.nature.com/articles/s42254-020-0195-3
10. H. Katori, M. Takamoto, V.G. Pal'chikov, V.D. Ovsiannikov, Phys. Rev. Lett. **91**(17), 173005 (2003). https://doi.org/10.1103/PhysRevLett.91.173005. https://journals.aps.org/prl/pdf/10.1103/PhysRevLett.91.173005
11. M. Takamoto, F.L. Hong, R. Higashi, H. Katori, Nature **435**(7040), 321 (2005). https://doi.org/10.1038/nature03541. https://www.nature.com/articles/nature03541
12. A.D. Ludlow, M.M. Boyd, J. Ye, E. Peik, P.O. Schmidt, Rev. Mod. Phys. **87**(2), 637 (2015). https://doi.org/10.1103/RevModPhys.87.637. https://link.aps.org/doi/10.1103/RevModPhys.87.637
13. A.M. Kaufman, K.K. Ni, Nat. Phys. **17**(12), 1324 (2021). https://doi.org/10.1038/s41567-021-01357-2. https://www.nature.com/articles/s41567-021-01357-2
14. N. Schlosser, G. Reymond, I. Protsenko, P. Grangier, Nature **411**(6841), 1024 (2001). https://doi.org/10.1038/35082512. http://dx.doi.org/10.1038/35082512
15. S. Kuhr, W. Alt, D. Schrader, M. Muller, V. Gomer, D. Meschede, Science **293**(5528), 278 (2001). https://doi.org/10.1126/science.1062725
16. A.M. Kaufman, B.J. Lester, C.A. Regal, Phys. Rev. X **2**(4), 041014 (2012). https://doi.org/10.1103/PhysRevX.2.041014. https://link.aps.org/doi/10.1103/PhysRevX.2.041014
17. J.D. Thompson, T.G. Tiecke, A.S. Zibrov, V. Vuletić, M.D. Lukin, Phys. Rev. Lett. **110**(13), 133001 (2013). https://doi.org/10.1103/PhysRevLett.110.133001

18. A.M. Kaufman, B.J. Lester, C.M. Reynolds, M.L. Wall, M. Foss-Feig, K.R.A. Hazzard, A.M. Rey, C.A. Regal, Science **345**(6194), 306 (2014). https://doi.org/10.1126/science.1250057
19. A.M. Kaufman, B.J. Lester, M. Foss-Feig, M.L. Wall, A.M. Rey, C.A. Regal, Nature **527**(7577), 208 (2015). https://doi.org/10.1038/nature16073
20. M.P.A. Jones, J. Beugnon, A. Gaëtan, J. Zhang, G. Messin, A. Browaeys, P. Grangier, Phys. Rev. A **75**(4), 040301 (2007). https://doi.org/10.1103/PhysRevA.75.040301. https://link.aps.org/doi/10.1103/PhysRevA.75.040301
21. J. Beugnon, C. Tuchendler, H. Marion, A. Gaëtan, Y. Miroshnychenko, Y.R.P. Sortais, A.M. Lance, M.P.A. Jones, G. Messin, A. Browaeys, P. Grangier, Nat. Phys. **3**(10), 696 (2007). https://doi.org/10.1038/nphys698. https://www.nature.com/articles/nphys698
22. L. Isenhower, E. Urban, X.L. Zhang, A.T. Gill, T. Henage, T.A. Johnson, T.G. Walker, M. Saffman, Phys. Rev. Lett. **104**(1), 010503 (2010). https://doi.org/10.1103/PhysRevLett.104.010503. http://link.aps.org/pdf/10.1103/PhysRevLett.104.010503
23. X.L. Zhang, L. Isenhower, A.T. Gill, T.G. Walker, M. Saffman, Phys. Rev. A **82**(3), 030306 (2010). https://doi.org/10.1103/PhysRevA.82.030306. https://link.aps.org/doi/10.1103/PhysRevA.82.030306
24. T. Wilk, A. Gaëtan, C. Evellin, J. Wolters, Y. Miroshnychenko, P. Grangier, A. Browaeys, Phys. Rev. Lett. **104**(1), 010502 (2010). https://doi.org/10.1103/PhysRevLett.104.010502. https://journals.aps.org/prl/pdf/10.1103/PhysRevLett.104.010502
25. K.M. Maller, M.T. Lichtman, T. Xia, Y. Sun, M.J. Piotrowicz, A.W. Carr, L. Isenhower, M. Saffman, Phys. Rev. A **92**(2), 022336 (2015). https://doi.org/10.1103/PhysRevA.92.022336. https://journals.aps.org/pra/pdf/10.1103/PhysRevA.92.022336
26. T. Xia, M. Lichtman, K. Maller, A.W. Carr, M.J. Piotrowicz, L. Isenhower, M. Saffman, Phys. Rev. Lett. **114**(10), 100503 (2015). https://doi.org/10.1103/PhysRevLett.114.100503
27. Y.Y. Jau, A.M. Hankin, T. Keating, I.H. Deutsch, G.W. Biedermann, Nat. Phys. **12**(1), 71 (2016). https://doi.org/10.1038/nphys3487. https://www.nature.com/articles/nphys3487
28. M. Endres, H. Bernien, A. Keesling, H. Levine, E.R. Anschuetz, A. Krajenbrink, C. Senko, V. Vuletic, M. Greiner, M.D. Lukin, Science **354**(6315), 1024 (2016). https://doi.org/10.1126/science.aah3752. https://www.science.org/doi/abs/10.1126/science.aah3752
29. D. Barredo, S. de Léséleuc, V. Lienhard, T. Lahaye, A. Browaeys, Science **354**(6315), 1021 (2016). https://doi.org/10.1126/science.aah3778
30. H. Kim, W. Lee, H.G. Lee, H. Jo, Y. Song, J. Ahn, Nat. Commun. **7**, 13317 (2016). https://doi.org/10.1038/ncomms13317
31. A. Browaeys, T. Lahaye, Nat. Phys. **16**(2), 132 (2020). https://doi.org/10.1038/s41567-019-0733-z. https://www.nature.com/articles/s41567-019-0733-z
32. H. Labuhn, D. Barredo, S. Ravets, S. de Léséleuc, T. Macrì, T. Lahaye, A. Browaeys, Nature **534**(7609), 667 (2016). https://doi.org/10.1038/nature18274
33. H. Bernien, S. Schwartz, A. Keesling, H. Levine, A. Omran, H. Pichler, S. Choi, A.S. Zibrov, M. Endres, M. Greiner, V. Vuletić, M.D. Lukin, Nature **551**(7682), 579 (2017). https://doi.org/10.1038/nature24622
34. V. Lienhard, S. de Léséleuc, D. Barredo, T. Lahaye, A. Browaeys, M. Schuler, L.P. Henry, A.M. Läuchli, Phys. Rev. X **8**(2), 021070 (2018). https://doi.org/10.1103/PhysRevX.8.021070. https://link.aps.org/doi/10.1103/PhysRevX.8.021070
35. A. Keesling, A. Omran, H. Levine, H. Bernien, H. Pichler, S. Choi, R. Samajdar, S. Schwartz, P. Silvi, S. Sachdev, P. Zoller, M. Endres, M. Greiner, V. Vuletić, M.D. Lukin, Nature **568**(7751), 207 (2019). https://doi.org/10.1038/s41586-019-1070-1. http://dx.doi.org/10.1038/s41586-019-1070-1
36. P. Scholl, M. Schuler, H.J. Williams, A.A. Eberharter, D. Barredo, K.N. Schymik, V. Lienhard, L.P. Henry, T.C. Lang, T. Lahaye, A.M. Läuchli, A. Browaeys, Nature **595**(7866), 233 (2021). https://doi.org/10.1038/s41586-021-03585-1. http://dx.doi.org/10.1038/s41586-021-03585-1
37. S. Ebadi, T.T. Wang, H. Levine, A. Keesling, G. Semeghini, A. Omran, D. Bluvstein, R. Samajdar, H. Pichler, W.W. Ho, S. Choi, S. Sachdev, M. Greiner, V. Vuletić, M.D. Lukin, Nature (London) **595**(7866), 227 (2021). https://doi.org/10.1038/s41586-021-03582-4. http://dx.doi.org/10.1038/s41586-021-03582-4

38. D. Bluvstein, A. Omran, H. Levine, A. Keesling, G. Semeghini, S. Ebadi, T.T. Wang, A.A. Michailidis, N. Maskara, W.W. Ho, S. Choi, M. Serbyn, M. Greiner, V. Vuletić, M.D. Lukin, Science **371**(6536), 1355 (2021). https://doi.org/10.1126/science.abg2530
39. P. Scholl, H.J. Williams, G. Bornet, F. Wallner, D. Barredo, L. Henriet, A. Signoles, C. Hainaut, T. Franz, S. Geier, A. Tebben, A. Salzinger, G. Zürn, T. Lahaye, M. Weidemüller, A. Browaeys, PRX Quantum **3**(2), 020303 (2022). https://doi.org/10.1103/PRXQuantum.3.020303. https://link.aps.org/doi/10.1103/PRXQuantum.3.020303
40. L.M. Steinert, P. Osterholz, R. Eberhard, L. Festa, N. Lorenz, Z. Chen, A. Trautmann, C. Gross, Phys. Rev. Lett. **130**(24), 243001 (2023). https://doi.org/10.1103/PhysRevLett.130.243001
41. C. Chen, G. Bornet, M. Bintz, G. Emperauger, L. Leclerc, V.S. Liu, P. Scholl, D. Barredo, J. Hauschild, S. Chatterjee, M. Schuler, A.M. Läuchli, M.P. Zaletel, T. Lahaye, N.Y. Yao, A. Browaeys, Nature **616**(7958), 691 (2023). https://doi.org/10.1038/s41586-023-05859-2
42. S. de Léséleuc, V. Lienhard, P. Scholl, D. Barredo, S. Weber, N. Lang, H.P. Büchler, T. Lahaye, A. Browaeys, Science **365**(6455), 775 (2019). https://doi.org/10.1126/science.aav9105. https:///www.science.org/doi/10.1126/science.aav9105?url_ver=Z39.88-2003&rfr_id=ori:rid:crossref.org&rfr_dat=cr_pub%3dpubmed
43. G. Semeghini, H. Levine, A. Keesling, S. Ebadi, T.T. Wang, D. Bluvstein, R. Verresen, H. Pichler, M. Kalinowski, R. Samajdar, A. Omran, S. Sachdev, A. Vishwanath, M. Greiner, V. Vuletić, M.D. Lukin, Science **374**(6572), 1242 (2021). https://doi.org/10.1126/science.abi8794
44. B.M. Spar, E. Guardado-Sanchez, S. Chi, Z.Z. Yan, W.S. Bakr, Phys. Rev. Lett. **128**(22), 223202 (2022). https://doi.org/10.1103/PhysRevLett.128.223202. https://link.aps.org/doi/10.1103/PhysRevLett.128.223202
45. A.W. Young, W.J. Eckner, N. Schine, A.M. Childs, A.M. Kaufman, Science **377**(6608), 885 (2022). https://doi.org/10.1126/science.abo0608
46. A. Omran, H. Levine, A. Keesling, G. Semeghini, T.T. Wang, S. Ebadi, H. Bernien, A.S. Zibrov, H. Pichler, S. Choi, J. Cui, M. Rossignolo, P. Rembold, S. Montangero, T. Calarco, M. Endres, M. Greiner, V. Vuletić, M.D. Lukin, Science **365**(6453), 570 (2019). https://doi.org/10.1126/science.aax9743
47. X. Wu, Z. Wang, F. Yang, R. Gao, C. Liang, M.K. Tey, X. Li, T. Pohl, L. You, Nat. Phys. **20**(9), 1389 (2024). https://doi.org/10.1038/s41567-024-02542-9. https://www.nature.com/articles/s41567-024-02542-9
48. B. Liu, L.H. Zhang, Z.K. Liu, J. Zhang, Z.Y. Zhang, S.Y. Shao, Q. Li, H.C. Chen, Y. Ma, T.Y. Han, Q.F. Wang, D.S. Ding, B.S. Shi (2024). arXiv:2402.13644
49. A.W. Young, S. Geller, W.J. Eckner, N. Schine, S. Glancy, E. Knill, A.M. Kaufman, Nature **629**(8011), 311 (2024). https://doi.org/10.1038/s41586-024-07304-4. https://www.nature.com/articles/s41586-024-07304-4
50. D. González-Cuadra, T.V. Zache, J. Carrasco, B. Kraus, P. Zoller, Phys. Rev. Lett. **129**(16), 160501 (2022). https://doi.org/10.1103/PhysRevLett.129.160501
51. M.A. Norcia, A.W. Young, W.J. Eckner, E. Oelker, J. Ye, A.M. Kaufman, Science **366**(6461), 93 (2019). https://doi.org/10.1126/science.aay0644
52. A.W. Young, W.J. Eckner, W.R. Milner, D. Kedar, M.A. Norcia, E. Oelker, N. Schine, J. Ye, A.M. Kaufman, Nature **588**(7838), 408 (2020). https://doi.org/10.1038/s41586-020-3009-y
53. A.L. Shaw, R. Finkelstein, R.B.S. Tsai, P. Scholl, T.H. Yoon, J. Choi, M. Endres, Nat. Phys. **20**(2), 195 (2024). https://doi.org/10.1038/s41567-023-02323-w. https://www.nature.com/articles/s41567-023-02323-w
54. A. Cao, W.J. Eckner, T. Lukin Yelin, A.W. Young, S. Jandura, L. Yan, K. Kim, G. Pupillo, J. Ye, N. Darkwah Oppong, A.M. Kaufman, Nature (London) **634**(8033), 315 (2024). https://doi.org/10.1038/s41586-024-07913-z. https://www.nature.com/articles/s41586-024-07913-z
55. W.J. Eckner, N. Darkwah Oppong, A. Cao, A.W. Young, W.R. Milner, J.M. Robinson, J. Ye, A.M. Kaufman, Nature 1–6 (2023). https://doi.org/10.1038/s41586-023-06360-6. https://www.nature.com/articles/s41586-023-06360-6
56. G. Bornet, G. Emperauger, C. Chen, B. Ye, M. Block, M. Bintz, J.A. Boyd, D. Barredo, T. Comparin, F. Mezzacapo, T. Roscilde, T. Lahaye, N.Y. Yao, A. Browaeys, Nature 1–6 (2023). https://doi.org/10.1038/s41586-023-06414-9. https://www.nature.com/articles/s41586-023-06414-9

57. M. Saffman, T.G. Walker, K. Mølmer, Rev. Mod. Phys. **82**(3), 2313 (2010). https://doi.org/10.1103/RevModPhys.82.2313. https://link.aps.org/doi/10.1103/RevModPhys.82.2313
58. M. Saffman, J. Phys. B: At. Mol. Opt. Phys. **49**(20), 202001 (2016). https://doi.org/10.1088/0953-4075/49/20/202001. https://iopscience.iop.org/article/10.1088/0953-4075/49/20/202001/meta
59. L. Henriet, L. Beguin, A. Signoles, T. Lahaye, A. Browaeys, G.O. Reymond, C. Jurczak, Quantum **4**, 327 (2020). https://doi.org/10.22331/q-2020-09-21-327. https://quantum-journal.org/papers/q-2020-09-21-327/pdf/
60. M. Morgado, S. Whitlock, AVS Quantum Sci. **3**(2), 023501 (2021). https://doi.org/10.1116/5.0036562
61. K. Wintersperger, F. Dommert, T. Ehmer, A. Hoursanov, J. Klepsch, W. Mauerer, G. Reuber, T. Strohm, M. Yin, S. Luber, EPJ Quantum Technology **10**(1), 1 (2023). https://doi.org/10.1140/epjqt/s40507-023-00190-1. https://epjquantumtechnology.springeropen.com/articles/10.1140/epjqt/s40507-023-00190-1
62. D. Ohl de Mello, D. Schäffner, J. Werkmann, T. Preuschoff, L. Kohfahl, M. Schlosser, G. Birkl, Phys. Rev. Lett. **122**(20), 203601 (2019). https://doi.org/10.1103/PhysRevLett.122.203601. https://link.aps.org/doi/10.1103/PhysRevLett.122.203601
63. P. Huft, Y. Song, T.M. Graham, K. Jooya, S. Deshpande, C. Fang, M. Kats, M. Saffman, Phys. Rev. A **105**(6), 063111 (2022). https://doi.org/10.1103/PhysRevA.105.063111. https://link.aps.org/doi/10.1103/PhysRevA.105.063111
64. K.N. Schymik, B. Ximenez, E. Bloch, D. Dreon, A. Signoles, F. Nogrette, D. Barredo, A. Browaeys, T. Lahaye, Phys. Rev. A (Coll. Park.) **106**(2), 022611 (2022). https://doi.org/10.1103/physreva.106.022611. https://link.aps.org/doi/10.1103/PhysRevA.106.022611
65. R. Tao, M. Ammenwerth, F. Gyger, I. Bloch, J. Zeiher, Phys. Rev. Lett. **133**(1), 013401 (2024). https://doi.org/10.1103/physrevlett.133.013401
66. F. Gyger, M. Ammenwerth, R. Tao, H. Timme, S. Snigirev, I. Bloch, J. Zeiher, Phys. Rev. Res. **6**(3), 033104 (2024). https://doi.org/10.1103/physrevresearch.6.033104
67. L. Pause, L. Sturm, M. Mittenbühler, S. Amann, T. Preuschoff, D. Schäffner, M. Schlosser, G. Birkl, Optica **11**(2), 222 (2024). https://doi.org/10.1364/OPTICA.513551. https://opg.optica.org/abstract.cfm?uri=optica-11-2-222
68. M.A. Norcia, H. Kim, W.B. Cairncross, M. Stone, A. Ryou, M. Jaffe, M.O. Brown, K. Barnes, P. Battaglino, T.C. Bohdanowicz, A. Brown, K. Cassella, C.A. Chen, R. Coxe, D. Crow, J. Epstein, C. Griger, E. Halperin, F. Hummel, A.M.W. Jones, J.M. Kindem, J. King, K. Kotru, J. Lauigan, M. Li, M. Lu, E. Megidish, J. Marjanovic, M. McDonald, T. Mittiga, J.A. Muniz, S. Narayanaswami, C. Nishiguchi, T. Paule, K.A. Pawlak, L.S. Peng, K.L. Pudenz, D. Rodríguez Pérez, A. Smull, D. Stack, M. Urbanek, R.J.M. van de Veerdonk, Z. Vendeiro, L. Wadleigh, T. Wilkason, T.Y. Wu, X. Xie, E. Zalys-Geller, X. Zhang, B.J. Bloom, PRX quantum **5**(3), 030316 (2024). https://doi.org/10.1103/prxquantum.5.030316
69. H.J. Manetsch, G. Nomura, E. Bataille, K.H. Leung, X. Lv, M. Endres (2024). arXiv:2403.12021
70. G. Pichard, D. Lim, E. Bloch, J. Vaneecloo, L. Bourachot, G.J. Both, G. Mériaux, S. Dutartre, R. Hostein, J. Paris, B. Ximenez, A. Signoles, A. Browaeys, T. Lahaye, D. Dreon, Phys. Rev. Appl. **22**(2), 024073 (2024). https://doi.org/10.1103/physrevapplied.22.024073
71. L. Rui, Z. Han-Sen, L. You, Z. Zhang-Rui, Z. Le-Tian, H. Tai-Ran, W. Hong-Ming, W. Zhan, M. Wei-Jie, G. Yan, Z. Yi-Kang, S. Zhao-Feng, O. Wan-Li, Z. Yu-Chen, R. Jun, C. Ming-Cheng, L. Chao-Yang, P. Jian-Wei (2024). arXiv:2412.14647
72. H. Levine, A. Keesling, A. Omran, H. Bernien, S. Schwartz, A.S. Zibrov, M. Endres, M. Greiner, V. Vuletić, M.D. Lukin, Phys. Rev. Lett. **121**(12), 123603 (2018). https://doi.org/10.1103/PhysRevLett.121.123603
73. H. Levine, A. Keesling, G. Semeghini, A. Omran, T.T. Wang, S. Ebadi, H. Bernien, M. Greiner, V. Vuletić, H. Pichler, M.D. Lukin, Phys. Rev. Lett. **123**(17), 170503 (2019). https://doi.org/10.1103/PhysRevLett.123.170503
74. I.S. Madjarov, J.P. Covey, A.L. Shaw, J. Choi, A. Kale, A. Cooper, H. Pichler, V. Schkolnik, J.R. Williams, M. Endres, Nat. Phys. **16**(8), 857 (2020). https://doi.org/10.1038/s41567-020-0903-z. https://www.nature.com/articles/s41567-020-0903-z

75. S. Ma, G. Liu, P. Peng, B. Zhang, S. Jandura, J. Claes, A.P. Burgers, G. Pupillo, S. Puri, J.D. Thompson, Nature (London) **622**(7982), 279 (2023). https://doi.org/10.1038/s41586-023-06438-1. http://dx.doi.org/10.1038/s41586-023-06438-1
76. S.J. Evered, D. Bluvstein, M. Kalinowski, S. Ebadi, T. Manovitz, H. Zhou, S.H. Li, A.A. Geim, T.T. Wang, N. Maskara, H. Levine, G. Semeghini, M. Greiner, V. Vuletić, M.D. Lukin, Nature (London) **622**(7982), 268 (2023). https://doi.org/10.1038/s41586-023-06481-y
77. P. Scholl, A.L. Shaw, R.B.S. Tsai, R. Finkelstein, J. Choi, M. Endres, Nature (London) **622**(7982), 273 (2023). https://doi.org/10.1038/s41586-023-06516-4. https://www.nature.com/articles/s41586-023-06516-4
78. M. Peper, Y. Li, D.Y. Knapp, M. Bileska, S. Ma, G. Liu, P. Peng, B. Zhang, S.P. Horvath, A.P. Burgers, J.D. Thompson, Phys. Rev. X. **15**(1), 011009 (2025). https://doi.org/10.1103/physrevx.15.011009. http://link.aps.org/pdf/10.1103/PhysRevX.15.011009
79. R.B.S. Tsai, X. Sun, A.L. Shaw, R. Finkelstein, M. Endres, PRX quantum **6**(1), 010331 (2025). https://doi.org/10.1103/prxquantum.6.010331. http://link.aps.org/pdf/10.1103/PRXQuantum.6.010331
80. A.G. Radnaev, W.C. Chung, D.C. Cole, D. Mason, T.G. Ballance, M.J. Bedalov, D.A. Belknap, M.R. Berman, M. Blakely, I.L. Bloomfield, P.D. Buttler, C. Campbell, A. Chopinaud, E. Copenhaver, M.K. Dawes, S.Y. Eubanks, A.J. Friss, D.M. Garcia, J. Gilbert, M. Gillette, P. Goiporia, P. Gokhale, J. Goldwin, D. Goodwin, T.M. Graham, C.J. Guttormsson, G.T. Hickman, L. Hurtley, M. Iliev, E.B. Jones, R.A. Jones, K.W. Kuper, T.B. Lewis, M.T. Lichtman, F. Majdeteimouri, J.J. Mason, J.K. McMaster, J.A. Miles, P.T. Mitchell, J.D. Murphree, N.A. Neff-Mallon, T. Oh, V. Omole, C.P. Simon, N. Pederson, M.A. Perlin, A. Reiter, R. Rines, P. Romlow, A.M. Scott, D. Stiefvater, J.R. Tanner, A.K. Tucker, I.V. Vinogradov, M.L. Warter, M. Yeo, M. Saffman, T.W. Noel (2024). arXiv:2408.08288
81. J.A. Muniz, M. Stone, D.T. Stack, M. Jaffe, J.M. Kindem, L. Wadleigh, E. Zalys-Geller, X. Zhang, C.A. Chen, M.A. Norcia, J. Epstein, E. Halperin, F. Hummel, T. Wilkason, M. Li, K. Barnes, P. Battaglino, T.C. Bohdanowicz, G. Booth, A. Brown, M.O. Brown, W.B. Cairncross, K. Cassella, R. Coxe, D. Crow, M. Feldkamp, C. Griger, A. Heinz, A.M.W. Jones, H. Kim, J. King, K. Kotru, J. Lauigan, J. Marjanovic, E. Megidish, M. Meredith, M. McDonald, R. Morshead, S. Narayanaswami, C. Nishiguchi, T. Paule, K.A. Pawlak, K.L. Pudenz, D.R. Pérez, A. Ryou, J. Simon, A. Smull, M. Urbanek, R.J.M. van de Veerdonk, Z. Vendeiro, T.Y. Wu, X. Xie, B.J. Bloom (2024). arXiv:2411.11708
82. L. Egan, D.M. Debroy, C. Noel, A. Risinger, D. Zhu, D. Biswas, M. Newman, M. Li, K.R. Brown, M. Cetina, C. Monroe, Nature (London) **598**(7880), 281 (2021). https://doi.org/10.1038/s41586-021-03928-y
83. M.P. da Silva, C. Ryan-Anderson, J.M. Bello-Rivas, A. Chernoguzov, J.M. Dreiling, C. Foltz, F. Frachon, J.P. Gaebler, T.M. Gatterman, L. Grans-Samuelsson, D. Hayes, N. Hewitt, J. Johansen, D. Lucchetti, M. Mills, S.A. Moses, B. Neyenhuis, A. Paz, J. Pino, P. Siegfried, J. Strabley, A. Sundaram, D. Tom, S.J. Wernli, M. Zanner, R.P. Stutz, K.M. Svore (2024). arXiv:2404.02280
84. Google Quantum AI, Nature (London) **595**(7867), 383 (2021). https://doi.org/10.1038/s41586-021-03588-y
85. Y. Zhao, Y. Ye, H.L. Huang, Y. Zhang, D. Wu, H. Guan, Q. Zhu, Z. Wei, T. He, S. Cao, F. Chen, T.H. Chung, H. Deng, D. Fan, M. Gong, C. Guo, S. Guo, L. Han, N. Li, S. Li, Y. Li, F. Liang, J. Lin, H. Qian, H. Rong, H. Su, L. Sun, S. Wang, Y. Wu, Y. Xu, C. Ying, J. Yu, C. Zha, K. Zhang, Y.H. Huo, C.Y. Lu, C.Z. Peng, X. Zhu, J.W. Pan, Phys. Rev. Lett. **129**(3), 030501 (2022). https://doi.org/10.1103/PhysRevLett.129.030501
86. S. Krinner, N. Lacroix, A. Remm, A. Di Paolo, E. Genois, C. Leroux, C. Hellings, S. Lazar, F. Swiadek, J. Herrmann, G.J. Norris, C.K. Andersen, M. Müller, A. Blais, C. Eichler, A. Wallraff, Nature (London) **605**(7911), 669 (2022). https://doi.org/10.1038/s41586-022-04566-8. https://www.nature.com/articles/s41586-022-04566-8
87. Google Quantum AI and Collaborators, Nature **638**(8052), 920 (2025). https://doi.org/10.1038/s41586-024-08449-y

88. D. Bluvstein, H. Levine, G. Semeghini, T.T. Wang, S. Ebadi, M. Kalinowski, A. Keesling, N. Maskara, H. Pichler, M. Greiner, V. Vuletić, M.D. Lukin, Nature (London) **604**(7906), 451 (2022). https://doi.org/10.1038/s41586-022-04592-6. https://www.nature.com/articles/s41586-022-04592-6
89. D. Bluvstein, S.J. Evered, A.A. Geim, S.H. Li, H. Zhou, T. Manovitz, S. Ebadi, M. Cain, M. Kalinowski, D. Hangleiter, J.P. Bonilla Ataides, N. Maskara, I. Cong, X. Gao, P. Sales Rodriguez, T. Karolyshyn, G. Semeghini, M.J. Gullans, M. Greiner, V. Vuletić, M.D. Lukin, Nature (London) **626**(7997), 58 (2024). https://doi.org/10.1038/s41586-023-06927-3. http://dx.doi.org/10.1038/s41586-023-06927-3
90. Q. Xu, J.P. Bonilla Ataides, C.A. Pattison, N. Raveendran, D. Bluvstein, J. Wurtz, B. Vasić, M.D. Lukin, L. Jiang, H. Zhou, Nat. Phys. **20**(7), 1084 (2024). https://doi.org/10.1038/s41567-024-02479-z. https://www.nature.com/articles/s41567-024-02479-z
91. G. Anqi, M.R. Joseph (2024). arXiv:2410.23263
92. C. Zi-Han, C. Ming-Cheng, L. Chao-Yang, P. Jian-Wei (2024). arXiv:2412.01391
93. B.W. Reichardt, A. Paetznick, D. Aasen, I. Basov, J.M. Bello-Rivas, P. Bonderson, R. Chao, W. van Dam, M.B. Hastings, A. Paz, M.P. da Silva, A. Sundaram, K.M. Svore, A. Vaschillo, Z. Wang, M. Zanner, W.B. Cairncross, C.A. Chen, D. Crow, H. Kim, J.M. Kindem, J. King, M. McDonald, M.A. Norcia, A. Ryou, M. Stone, L. Wadleigh, K. Barnes, P. Battaglino, T.C. Bohdanowicz, G. Booth, A. Brown, M.O. Brown, K. Cassella, R. Coxe, J.M. Epstein, M. Feldkamp, C. Griger, E. Halperin, A. Heinz, F. Hummel, M. Jaffe, A.M.W. Jones, E. Kapit, K. Kotru, J. Lauigan, M. Li, J. Marjanovic, E. Megidish, M. Meredith, R. Morshead, J.A. Muniz, S. Narayanaswami, C. Nishiguchi, T. Paule, K.A. Pawlak, K.L. Pudenz, D.R. Pérez, J. Simon, A. Smull, D. Stack, M. Urbanek, R.J.M. van de Veerdonk, Z. Vendeiro, R.T. Weverka, T. Wilkason, T.Y. Wu, X. Xie, E. Zalys-Geller, X. Zhang, B.J. Bloom (2024). arXiv:2411.11822
94. M.J. Bedalov, M. Blakely, P.D. Buttler, C. Carnahan, F.T. Chong, W.C. Chung, D.C. Cole, P. Goiporia, P. Gokhale, B. Heim, G.T. Hickman, E.B. Jones, R.A. Jones, P. Khalate, J.S. Kim, K.W. Kuper, M.T. Lichtman, S. Lee, D. Mason, N.A. Neff-Mallon, T.W. Noel, V. Omole, A.G. Radnaev, R. Rines, M. Saffman, E. Shabtai, M.H. Teo, B. Thotakura, T. Tomesh, A.K. Tucker (2024). arXiv:2412.07670
95. F. Nogrette, H. Labuhn, S. Ravets, D. Barredo, L. Béguin, A. Vernier, T. Lahaye, A. Browaeys, Phys. Rev. X. **4**(2), 021034 (2014). https://doi.org/10.1103/physrevx.4.021034. https://link.aps.org/doi/10.1103/PhysRevX.4.021034
96. D. Barredo, V. Lienhard, S. de Léséleuc, T. Lahaye, A. Browaeys, Nature **561**(7721), 79 (2018). https://doi.org/10.1038/s41586-018-0450-2. https://www.nature.com/articles/s41586-018-0450-2
97. M.O. Brown, T. Thiele, C. Kiehl, T.W. Hsu, C.A. Regal, Phys. Rev. X **9**(1), 011057 (2019). https://doi.org/10.1103/PhysRevX.9.011057. https://link.aps.org/doi/10.1103/PhysRevX.9.011057
98. Z.Z. Yan, B.M. Spar, M.L. Prichard, S. Chi, H.T. Wei, E. Ibarra-García-Padilla, K.R.A. Hazzard, W.S. Bakr, Phys. Rev. Lett. **129**(12), 123201 (2022). https://doi.org/10.1103/PhysRevLett.129.123201.https://link.aps.org/doi/10.1103/PhysRevLett.129.123201
99. K.N. Blodgett, D. Peana, S.S. Phatak, L.M. Terry, M.P. Montes, J.D. Hood, Phys. Rev. Lett. **131**(8), 083001 (2023). https://doi.org/10.1103/PhysRevLett.131.083001
100. N.R. Hutzler, L.R. Liu, Y. Yu, K.K. Ni, New J. Phys. **19**(2), 023007 (2017). https://doi.org/10.1088/1367-2630/aa5a3b. https://iopscience.iop.org/article/10.1088/1367-2630/aa5a3b/meta
101. Y. Yu, N.R. Hutzler, J.T. Zhang, L.R. Liu, J.D. Hood, T. Rosenband, K.K. Ni, Phys. Rev. A **97**(6), 063423 (2018). https://doi.org/10.1103/PhysRevA.97.063423. https://link.aps.org/doi/10.1103/PhysRevA.97.063423
102. M.M. Aliyu, L. Zhao, X.Q. Quek, K.C. Yellapragada, H. Loh, Phys. Rev. Res. **3**(4) (2021). https://doi.org/10.1103/physrevresearch.3.043059. https://journals.aps.org/prresearch/abstract/10.1103/PhysRevResearch.3.043059

103. Y. Wang, S. Shevate, T.M. Wintermantel, M. Morgado, G. Lochead, S. Whitlock, Npj Quantum Inf. **6**(1), 1 (2020). https://doi.org/10.1038/s41534-020-0285-1. https://www.nature.com/articles/s41534-020-0285-1
104. N. Lorenz, L. Festa, L.M. Steinert, C. Gross, SciPost Phys. **10**(3), 052 (2021). https://doi.org/10.21468/scipostphys.10.3.052. https://scipost.org/10.21468/SciPostPhys.10.3.052/pdf
105. A. Cooper, J.P. Covey, I.S. Madjarov, S.G. Porsev, M.S. Safronova, M. Endres, Phys. Rev. X **8**(4), 041055 (2018). https://doi.org/10.1103/PhysRevX.8.041055. https://link.aps.org/doi/10.1103/PhysRevX.8.041055
106. M.A. Norcia, A.W. Young, A.M. Kaufman, Phys. Rev. X **8**(4), 041054 (2018). https://doi.org/10.1103/PhysRevX.8.041054. https://link.aps.org/doi/10.1103/PhysRevX.8.041054
107. K. Barnes, P. Battaglino, B.J. Bloom, K. Cassella, R. Coxe, N. Crisosto, J.P. King, S.S. Kondov, K. Kotru, S.C. Larsen, J. Lauigan, B.J. Lester, M. McDonald, E. Megidish, S. Narayanaswami, C. Nishiguchi, R. Notermans, L.S. Peng, A. Ryou, T.Y. Wu, M. Yarwood, Nat. Commun. **13**(1), 2779 (2022). https://doi.org/10.1038/s41467-022-29977-z. https://www.nature.com/articles/s41467-022-29977-z
108. A. Urech, I.H.A. Knottnerus, R.J.C. Spreeuw, F. Schreck, Phys. Rev. Research **4**(2), 023245 (2022). https://doi.org/10.1103/PhysRevResearch.4.023245. https://link.aps.org/doi/10.1103/PhysRevResearch.4.023245
109. G. Unnikrishnan, P. Ilzhöfer, A. Scholz, C. Hölzl, A. Götzelmann, R.K. Gupta, J. Zhao, J. Krauter, S. Weber, N. Makki, H.P. Büchler, T. Pfau, F. Meinert, Phys. Rev. Lett. **132**(15), 150606 (2024). https://doi.org/10.1103/PhysRevLett.132.150606. http://dx.doi.org/10.1103/PhysRevLett.132.150606
110. X. Yan, C. He, K. Wen, Z. Ren, P.T.F. Wong, E. Hajiyev, G.B. Jo, Phys. Rev. Res. **7**(1), 013031 (2025). https://doi.org/10.1103/physrevresearch.7.013031. http://dx.doi.org/10.1103/PhysRevResearch.7.013031
111. S. Saskin, J.T. Wilson, B. Grinkemeyer, J.D. Thompson, Phys. Rev. Lett. **122**(14), 143002 (2019). https://doi.org/10.1103/PhysRevLett.122.143002. http://dx.doi.org/10.1103/PhysRevLett.122.143002
112. D. Okuno, Y. Nakamura, T. Kusano, Y. Takasu, N. Takei, H. Konishi, Y. Takahashi, J. Phys. Soc. Jpn. **91**(8), 084301 (2022). https://doi.org/10.7566/JPSJ.91.084301
113. A. Jenkins, J.W. Lis, A. Senoo, W.F. McGrew, A.M. Kaufman, Phys. Rev. X **12**(2), 021027 (2022). https://doi.org/10.1103/PhysRevX.12.021027. https://link.aps.org/doi/10.1103/PhysRevX.12.021027
114. W. Huie, L. Li, N. Chen, X. Hu, Z. Jia, W.K.C. Sun, J.P. Covey, PRX Quantum **4**(3), 030337 (2023). https://doi.org/10.1103/PRXQuantum.4.030337. https://link.aps.org/doi/10.1103/PRXQuantum.4.030337
115. M.A. Norcia, W.B. Cairncross, K. Barnes, P. Battaglino, A. Brown, M.O. Brown, K. Cassella, C.A. Chen, R. Coxe, D. Crow, J. Epstein, C. Griger, A.M.W. Jones, H. Kim, J.M. Kindem, J. King, S.S. Kondov, K. Kotru, J. Lauigan, M. Li, M. Lu, E. Megidish, J. Marjanovic, M. McDonald, T. Mittiga, J.A. Muniz, S. Narayanaswami, C. Nishiguchi, R. Notermans, T. Paule, K.A. Pawlak, L.S. Peng, A. Ryou, A. Smull, D. Stack, M. Stone, A. Sucich, M. Urbanek, R.J.M. van de Veerdonk, Z. Vendeiro, T. Wilkason, T.Y. Wu, X. Xie, X. Zhang, B.J. Bloom, Phys. Rev. X **13**(4), 041034 (2023). https://doi.org/10.1103/PhysRevX.13.041034. https://link.aps.org/doi/10.1103/PhysRevX.13.041034
116. S. Ma, A.P. Burgers, G. Liu, J. Wilson, B. Zhang, J.D. Thompson, Phys. Rev. X **12**(2), 021028 (2022). https://doi.org/10.1103/PhysRevX.12.021028. https://link.aps.org/doi/10.1103/PhysRevX.12.021028
117. N. Schine, A.W. Young, W.J. Eckner, M.J. Martin, A.M. Kaufman, Nat. Phys. **18**(9), 1067 (2022). https://doi.org/10.1038/s41567-022-01678-w. https://www.nature.com/articles/s41567-022-01678-w
118. D.T.C. Allcock, W.C. Campbell, J. Chiaverini, I.L. Chuang, E.R. Hudson, I.D. Moore, A. Ransford, C. Roman, J.M. Sage, D.J. Wineland, Appl. Phys. Lett. **119**(21), 214002 (2021). https://doi.org/10.1063/5.0069544

119. J.W. Lis, A. Senoo, W.F. McGrew, F. Rönchen, A. Jenkins, A.M. Kaufman, Phys. Rev. X **13**(4), 041035 (2023). https://doi.org/10.1103/PhysRevX.13.041035. https://link.aps.org/doi/10.1103/PhysRevX.13.041035
120. J.T. Wilson, S. Saskin, Y. Meng, S. Ma, R. Dilip, A.P. Burgers, J.D. Thompson, Phys. Rev. Lett. **128**(3), 033201 (2022). https://doi.org/10.1103/PhysRevLett.128.033201
121. K.L. Pham, T.F. Gallagher, P. Pillet, S. Lepoutre, P. Cheinet, PRX Quantum **3**(2), 020327 (2022). https://doi.org/10.1103/PRXQuantum.3.020327. https://link.aps.org/doi/10.1103/PRXQuantum.3.020327
122. A.P. Burgers, S. Ma, S. Saskin, J. Wilson, M.A. Alarcón, C.H. Greene, J.D. Thompson, PRX Quantum **3**(2), 020326 (2022). https://doi.org/10.1103/PRXQuantum.3.020326. https://link.aps.org/doi/10.1103/PRXQuantum.3.020326
123. Y. Wu, S. Kolkowitz, S. Puri, J.D. Thompson, Nat. Commun. **13**(1), 4657 (2022). https://doi.org/10.1038/s41467-022-32094-6. https://www.nature.com/articles/s41467-022-32094-6

Chapter 2
Atom-Light Interaction

In this chapter, we review the interaction between atoms and light. First, we discuss multipole transitions, which are essential for controlling qubits. Next, we introduce atomic polarizability and the dipole potential, which are crucial for atom trapping. Finally, we formulate the dynamics of both the internal and external degrees of freedom of atoms under light irradiation.

2.1 Multipole Transitions

In this section, we review the multipole transitions between atoms and light, based on Refs. [1–3]. We consider the interaction between an atom and an electromagnetic field. The total Hamiltonian for a valence electron of an atom in an electromagnetic field $\boldsymbol{A}(\boldsymbol{r}, t)$ in the Coulomb gauge ($\nabla \cdot \boldsymbol{A} = 0$) is given by [4]

$$\mathcal{H} = \frac{1}{2m}\left(\boldsymbol{p} + e\boldsymbol{A}\right)^2 + V(\boldsymbol{r}) + \frac{\epsilon_0}{2}\int \mathrm{d}^3r\,(\boldsymbol{E}_{\perp}{}^2 + c^2\boldsymbol{B}^2), \tag{2.1}$$

where $e,\ m,\ \boldsymbol{r},\ \boldsymbol{p}$ is the charge, mass, position, and momentum of the valence electron, respectively. $V(\boldsymbol{r})$ is the potential energy which the electron feels from the other charged particles in the atom, and $\epsilon_0,\ c,\ \boldsymbol{E}_{\perp} = -\partial_t \boldsymbol{A},\ \boldsymbol{B} = \nabla \times \boldsymbol{A}$ is the electric permittivity in vacuum, the speed of light, the transverse component of the electric field and the magnetic field, respectively. The interaction part of the Hamiltonian arises from the first term in Eq. (2.1) as

$$\mathcal{H}_{\text{int}} = \frac{e}{m}\boldsymbol{p} \cdot \boldsymbol{A} + \frac{e^2}{2m}\boldsymbol{A}^2 \simeq \frac{e}{m}\boldsymbol{p} \cdot \boldsymbol{A}. \tag{2.2}$$

Y. Nakamura, *Generation of Highly Controllable Ytterbium Atom Array*,
Springer Theses, https://doi.org/10.1007/978-981-95-2836-3_2

Here, we have neglected the second-order term of $\boldsymbol{A}$ and used the fact that $[\boldsymbol{p}, \boldsymbol{A}] = 0$ in the Coulomb gauge. Additionally, we take into account the Zeeman interaction between the magnetic moment of the electron and the magnetic field, which is given by

$$\mathcal{H}_{\text{Zeeman}} = -\boldsymbol{\mu} \cdot \boldsymbol{B} = -\frac{e g_s}{2m} \boldsymbol{S} \cdot [\nabla \times \boldsymbol{A}], \tag{2.3}$$

where $\boldsymbol{\mu}$ is the magnetic moment of the electron, $\boldsymbol{S}$ is the spin operator, and g_s is the Landé g-factor. The total interaction Hamiltonian is then given by

$$\mathcal{H}_{\text{int}} = \frac{e}{m} \boldsymbol{p} \cdot \boldsymbol{A} - \frac{g_s e}{2m} \boldsymbol{S} \cdot [\nabla \times \boldsymbol{A}]. \tag{2.4}$$

The most common approach for derivation of the multipole transitions is to use Taylor expansion of the vector potential $\boldsymbol{A}(\boldsymbol{r}, t) = A_0 \hat{\boldsymbol{\epsilon}} \exp(i\boldsymbol{k} \cdot \boldsymbol{r} - i\omega t)$. Here, A_0 is the amplitude of the vector potential, $\boldsymbol{k}$ is the wave vector, ω is the frequency of the field, and $\hat{\boldsymbol{\epsilon}}$ is the polarization vector. In the Coulomb gauge, the electric field is given by

$$\begin{aligned} \boldsymbol{E}(\boldsymbol{r}, t) = -\partial_t \boldsymbol{A}(\boldsymbol{r}, t) &= i\omega A_0 \hat{\boldsymbol{\epsilon}} \exp(i\boldsymbol{k} \cdot \boldsymbol{r} - i\omega t) \\ &= i E_0 \hat{\boldsymbol{\epsilon}} \exp(i\boldsymbol{k} \cdot \boldsymbol{r} - i\omega t), \end{aligned} \tag{2.5}$$

where $E_0 = \omega A_0$. Assuming that the wavelength of the light is much larger than the size of the atom ($\boldsymbol{k} \cdot \boldsymbol{r} \ll 1$), we can expand the vector potential as

$$\boldsymbol{A}(\boldsymbol{r}, t) = A_0 \hat{\boldsymbol{\epsilon}} \left(1 + i\boldsymbol{k} \cdot \boldsymbol{r} - \frac{1}{2}(\boldsymbol{k} \cdot \boldsymbol{r})^2 + \cdots \right) \exp(-i\omega t). \tag{2.6}$$

The first term corresponds to the electric dipole (E1) transition, the second term to the electric quadrupole (E2) and magnetic dipole (M1) transitions, the third term to the electric octupole (E3) and magnetic quadrupole (M2) transitions, and so on.

As an example, we consider the electric dipole transition with this approach. The lowest order term of the interaction Hamiltonian in Eq. (2.4) becomes

$$\mathcal{H}_{\text{int}} = \frac{e}{m} \boldsymbol{p} \cdot \boldsymbol{A} \simeq i \frac{e E_0}{m\omega} \boldsymbol{p} \cdot \hat{\boldsymbol{\epsilon}} \exp(-i\omega t). \tag{2.7}$$

If the oscillation frequency of the light (ω) is close to the frequency of an atomic transition $|i\rangle \leftrightarrow |f\rangle$ (ω_{fi}), the coupling between the light and the atom is written as

$$i \frac{e E_0}{m\omega_{fi}} \langle f|\boldsymbol{p}|i\rangle \cdot \hat{\boldsymbol{\epsilon}} \exp(-i\omega_{fi} t). \tag{2.8}$$

We then define the coupling strength of the transition as

$$d_{fi} = \frac{1}{m\omega_{fi}} \langle f|\boldsymbol{p}|i\rangle \cdot \hat{\boldsymbol{\epsilon}}, \tag{2.9}$$

and recall that the commutation relation between the momentum operator $\boldsymbol{p}$ and the bare atomic Hamiltonian $\mathcal{H}_0 = \boldsymbol{p}^2/(2m)$ is written as $[\boldsymbol{r}, \mathcal{H}_0] = (i\hbar/m)\boldsymbol{p}$, where $\hbar = h/(2\pi)$ with the Planck constant h, the coupling strength can be rewritten as

$$\begin{aligned} d_{fi} &= \frac{1}{m\omega_{fi}} \langle f|\boldsymbol{p}|i\rangle \cdot \hat{\boldsymbol{\epsilon}} = \frac{1}{i\hbar\omega_{fi}} \langle f|[\boldsymbol{r}, \mathcal{H}_0]|i\rangle \cdot \hat{\boldsymbol{\epsilon}} \\ &= \frac{1}{i\hbar\omega_{fi}} (\langle f|\boldsymbol{r}\mathcal{H}_0|i\rangle - \langle f|\mathcal{H}_0\boldsymbol{r}|i\rangle) \cdot \hat{\boldsymbol{\epsilon}} \\ &= \frac{i}{\hbar} \langle f|\boldsymbol{r}|i\rangle \cdot \hat{\boldsymbol{\epsilon}}. \end{aligned} \tag{2.10}$$

Then, the coupling between the light and the atom (Eq. (2.8)) becomes

$$\begin{aligned} i\frac{eE_0}{m\omega_{fi}} \langle f|\boldsymbol{p}|i\rangle \cdot \hat{\boldsymbol{\epsilon}} \exp(-i\omega_{fi}t) &= ieE_0\frac{i}{\hbar} \langle f|\boldsymbol{r}|i\rangle \cdot \hat{\boldsymbol{\epsilon}} \exp(-i\omega_{fi}t) \\ &= -\frac{\left\langle f\middle|\boldsymbol{d} \cdot \hat{\boldsymbol{\epsilon}}\middle|i\right\rangle E_0}{\hbar} \exp(-i\omega_{fi}t), \end{aligned} \tag{2.11}$$

where $\boldsymbol{d} = e\boldsymbol{r}$ is the electric dipole operator. The coupling strength $\Omega := -\left\langle f\middle|\boldsymbol{d} \cdot \hat{\boldsymbol{\epsilon}}\middle|i\right\rangle E_0/\hbar$ is so-called the Rabi frequency of the electric dipole transition.

We now consider the higher-order multipole transitions. Instead of using the Taylor expansion of the vector potential, we can use the vector spherical harmonics to expand the vector potential. The vector spherical harmonics $\boldsymbol{Y}_{K,l,q}$ are defined as

$$\boldsymbol{Y}_{K,l,q} = (-1)^{K-q}\sqrt{2K+1} \sum_{p=-1}^{1} \begin{pmatrix} K & 1 & l \\ -q & p & q-p \end{pmatrix} Y_l^{q-p}(\theta, \phi)\hat{\boldsymbol{e}}_p, \tag{2.12}$$

where the parentheses denote the Wigner 3-j symbol, $Y_l^q(\theta, \phi)$ is the spherical harmonics, and $\hat{\boldsymbol{e}}_p$ is the unit vector of the p direction in spherical coordinate. The vector potential is then expanded as

$$\boldsymbol{A} = \sum_{K,l,q} \boldsymbol{A}_{K,l,q}\boldsymbol{Y}_{K,l,q}. \tag{2.13}$$

Here, $K = 1, 2, \cdots$ represents the multipole order, where $K = 1$ corresponds to the dipole term, $K = 2$ to the quadrupole, and so on. The label l and q runs over $l = K-1, K, K+1$ and $|q| \leq l$, respectively. The expansion coefficients $A_{K,l,q}$ are given by

$$A_{K,l,q} = A_0 \int_0^{\pi} \mathrm{d}\theta \sin\theta \int_0^{2\pi} \mathrm{d}\phi \boldsymbol{Y}_{K,l,q}(\hat{\boldsymbol{r}}) \cdot \hat{\boldsymbol{\epsilon}} e^{i\boldsymbol{k}\cdot\boldsymbol{r}}. \tag{2.14}$$

Substituting the expansion of a plane wave in terms of the spherical Bessel function $j_l(x)$ and the spherical harmonics $Y_l^m(\theta, \phi)$,

$$e^{i\boldsymbol{k}\cdot\boldsymbol{r}} = 4\pi \sum_{l,q} i^l j_l(kr) Y_l^q(\hat{\boldsymbol{r}}) Y_l^{m*}(\hat{\boldsymbol{k}}) \tag{2.15}$$

with $\hat{\boldsymbol{r}} = \boldsymbol{r}/|\boldsymbol{r}|$ and $\hat{\boldsymbol{k}} = \boldsymbol{k}/|\boldsymbol{k}|$ into Eq. (2.14), we obtain

$$\boldsymbol{A}(\boldsymbol{r}) = 4\pi \sum_{K,l,q} i^l (\boldsymbol{Y}_{K,l,q}(\hat{\boldsymbol{k}}) \cdot \hat{\boldsymbol{\epsilon}}) j_l(kr) \boldsymbol{Y}_{K,l,q}(\hat{\boldsymbol{r}}). \tag{2.16}$$

Here we introduce another set of vector spherical harmonics $\boldsymbol{Y}^{(\mu)}_{K,q}$, which is defined as

$$\begin{aligned}
\boldsymbol{Y}^{(-1)}_{K,q}(\hat{\boldsymbol{r}}) &= \sqrt{\frac{K}{2K+1}}\boldsymbol{Y}_{K,K-1,q}(\hat{\boldsymbol{r}}) - \sqrt{\frac{K+1}{2K+1}}\boldsymbol{Y}_{K,K+1,q}(\hat{\boldsymbol{r}}) = \hat{\boldsymbol{r}} Y_K^q(\hat{\boldsymbol{r}}),\\
\boldsymbol{Y}^{(0)}_{K,q}(\hat{\boldsymbol{r}}) &= \boldsymbol{Y}_{K,K,q}(\hat{\boldsymbol{r}}) = \frac{1}{\sqrt{K(K+1)}}(-i\boldsymbol{r}\times\nabla) Y_K^q(\hat{\boldsymbol{r}}),\\
\boldsymbol{Y}^{(1)}_{K,q}(\hat{\boldsymbol{r}}) &= \sqrt{\frac{K+1}{2K+1}}\boldsymbol{Y}_{K,K-1,q}(\hat{\boldsymbol{r}}) + \sqrt{\frac{K}{2K+1}}\boldsymbol{Y}_{K,K+1,q}(\hat{\boldsymbol{r}}) = \frac{r}{\sqrt{K(K+1)}}\nabla Y_K^q(\hat{\boldsymbol{r}}).
\end{aligned} \tag{2.17}$$

Using these $\boldsymbol{Y}^{(\mu)}_{K,q}$, the expansion of the vector potential in Eq. (2.16) can be rewritten as

$$\boldsymbol{A}(\boldsymbol{r}) = 4\pi \sum_{K,q,\mu} i^{K-\mu} (\boldsymbol{Y}^{(\mu)}_{K,q}(\hat{\boldsymbol{k}}) \cdot \hat{\boldsymbol{\epsilon}}) \boldsymbol{a}^{(\mu)}_{K,q}(\hat{\boldsymbol{r}}), \tag{2.18}$$

with

$$\begin{aligned}
\boldsymbol{a}^{(0)}_{K,q}(\hat{\boldsymbol{r}}) &= A_0 j_K(kr) \boldsymbol{Y}^{(0)}_{K,q}(\hat{\boldsymbol{r}}),\\
\boldsymbol{a}^{(1)}_{K,q}(\hat{\boldsymbol{r}}) &= A_0 \left[\sqrt{\frac{K+1}{2K+1}} j_{K-1}(kr) \boldsymbol{Y}_{K,K-1,q}(\hat{\boldsymbol{r}}) - \sqrt{\frac{K}{2K+1}} j_{K+1}(kr) \boldsymbol{Y}_{K,K+1,q}(\hat{\boldsymbol{r}})\right].
\end{aligned} \tag{2.19}$$

Here, $\mu = 0$ and $\mu = 1$ corresponds to the magnetic and electric multipole components respectively. Note that since $\boldsymbol{Y}^{(-1)}_{K,q}(\hat{\boldsymbol{k}}) \propto \hat{\boldsymbol{k}}$, its contribution to the vector potential is zero under the assumption that the laser propagation axis is orthogonal to the polarization vector ($\boldsymbol{k} \cdot \hat{\boldsymbol{\epsilon}} = 0$). Substituting Eq. (2.19) into Eq. (2.4), the interaction Hamiltonian becomes

$$\begin{aligned}
\mathcal{H}_{\text{int}} &= 4\pi \sum_{\mu=0,1} \sum_{K=1}^{\infty} \sum_{q=-K}^{K} i^{K-\mu} (\boldsymbol{Y}^{(\mu)}_{K,q}(\hat{\boldsymbol{k}}) \cdot \hat{\boldsymbol{\epsilon}}) \left[\frac{e}{m}\boldsymbol{p} \cdot \boldsymbol{a}^{(\mu)}_{K,q}(\hat{\boldsymbol{r}}) + \frac{e g_s}{2m}\boldsymbol{S} \cdot (\nabla \times \boldsymbol{a}^{(\mu)}_{K,q}(\hat{\boldsymbol{r}}))\right]\\
&:= 4\pi \sum_{\mu=0,1} \sum_{K=1}^{\infty} \sum_{q=-K}^{K} i^{K-\mu} (\boldsymbol{Y}^{(\mu)}_{K,q}(\hat{\boldsymbol{k}}) \cdot \hat{\boldsymbol{\epsilon}}) \mathcal{H}^{(\mu)}_{K,q}.
\end{aligned} \tag{2.20}$$

2.1.1 Electric Multipole Transitions

The electric multipole component in the interaction Hamiltonian (Eq. (2.20)) is given by

$$\mathcal{H}_{K,q}^{(\mathrm{el})} := \mathcal{H}_{K,q}^{(1)} = \frac{e}{m}\boldsymbol{p}\cdot\boldsymbol{a}_{K,q}^{(1)}(\hat{\boldsymbol{r}}). \tag{2.21}$$

Note that since $\nabla\times\boldsymbol{a}_{K,q}^{(1)} \propto \nabla\times(\hat{\boldsymbol{r}} + \nabla Y_K^q) = 0$, the electric components does not contain the coupling to the spin operator $\boldsymbol{S}$. Using the Eqs. (2.17) and (2.20), the electric multipole component is rewritten as

$$\begin{aligned}
\boldsymbol{a}_{K,q}^{(1)}(\hat{\boldsymbol{r}}) &= A_0\left[\sqrt{\frac{K+1}{2K+1}}j_{K-1}(kr)\boldsymbol{Y}_{K,K-1,q}(\hat{\boldsymbol{r}}) - \sqrt{\frac{K}{2K+1}}j_{K+1}(kr)\boldsymbol{Y}_{K,K+1,q}(\hat{\boldsymbol{r}})\right]\\
&= A_0\frac{\sqrt{K(K+1)}}{2K+1}\boldsymbol{Y}_{K,q}^{(-1)}(\hat{\boldsymbol{r}})[j_{K-1}(kr) + j_{K+1}(kr)]\\
&\quad + A_0\frac{1}{2K+1}\boldsymbol{Y}_{K,q}^{(1)}(\hat{\boldsymbol{r}})[(K+1)j_{K-1}(kr) - Kj_{K+1}(kr)]\\
&= A_0\sqrt{K(K+1)}\frac{j_K(kr)}{kr}\boldsymbol{Y}_{K,q}^{(-1)}(\hat{\boldsymbol{r}}) + A_0\left[\frac{j_K(kr)}{kr} + \frac{\partial j_K(kr)}{\partial(kr)}\right]\boldsymbol{Y}_{K,q}^{(1)}(\hat{\boldsymbol{r}}).
\end{aligned} \tag{2.22}$$

Here we used the recurrence relation of the spherical Bessel function as

$$\begin{aligned}
j_{K-1}(kr) &= \frac{K+1}{kr}j_K(kr) + \frac{\partial j_K(kr)}{\partial(kr)},\\
j_{K+1}(kr) &= \frac{K}{kr}j_K(kr) - \frac{\partial j_K(kr)}{\partial(kr)}.
\end{aligned} \tag{2.23}$$

To further proceed the calculation, we introduce a gauge transformation of the vector potential as

$$\begin{aligned}
\boldsymbol{A}(\boldsymbol{r},t) &\to \boldsymbol{A}'(\boldsymbol{r},t) = \boldsymbol{A}(\boldsymbol{r},t) + \nabla\chi(\boldsymbol{r},t),\\
\Phi(\boldsymbol{r},t) &\to \Phi'(\boldsymbol{r},t) = \Phi(\boldsymbol{r},t) - \partial_t\chi(\boldsymbol{r},t),
\end{aligned} \tag{2.24}$$

with a gauge function $\chi(\boldsymbol{r},t)$ of

$$\chi(\boldsymbol{r},t) = -\frac{A_0}{k}\sqrt{\frac{K+1}{K}}j_K(kr)Y_K^q(\hat{\boldsymbol{r}})e^{i\omega t}. \tag{2.25}$$

This gauge transformation results in the following vector and scalar potentials as

$$\begin{aligned}
\boldsymbol{a}_{K,q}^{(1)}(\hat{\boldsymbol{r}}) &\to -A_0 j_{K+1}(kr)\left[\boldsymbol{Y}_{K,q}^{(1)}(\hat{\boldsymbol{r}}) - \sqrt{\frac{K+1}{K}}\boldsymbol{Y}_{K,q}^{(-1)}(\hat{\boldsymbol{r}})\right],\\
\Phi(\boldsymbol{r}) &\to -i\omega\chi(\boldsymbol{r},t),
\end{aligned} \tag{2.26}$$

and thus the interaction Hamiltonian of Eq. (2.21) becomes

$$
\begin{aligned}
\mathcal{H}_{K,q}^{(\mathrm{el})} &= \frac{e}{m}\boldsymbol{p}\cdot\boldsymbol{a}_{K,q}^{(1)}(\hat{\boldsymbol{r}}) + e\Phi(\boldsymbol{r}) \\
&= -\frac{eA_0}{m} j_{K+1}(kr)\boldsymbol{p}\cdot\left[\boldsymbol{Y}_{K,q}^{(1)}(\hat{\boldsymbol{r}}) - \sqrt{\frac{K+1}{K}}\boldsymbol{Y}_{K,q}^{(-1)}(\hat{\boldsymbol{r}})\right] + i\frac{\omega eA_0}{k}\sqrt{\frac{K+1}{K}} j_K(kr)Y_K^q(\hat{\boldsymbol{r}}).
\end{aligned}
\tag{2.27}
$$

By assuming the condition of $kr \ll 1$ and using the asymptotic form of the spherical Bessel function as $j_K(kr) \simeq (kr)^K/(2K+1)!!$, the lowest order term of the electric multipole Hamiltonian is given by

$$
\begin{aligned}
\mathcal{H}_{K,q}^{(\mathrm{el})} &\simeq i\frac{\omega eA_0}{k}\sqrt{\frac{K+1}{K}}\frac{(kr)^K}{(2K+1)!!}Y_K^q(\hat{\boldsymbol{r}}) \\
&= icA_0\sqrt{\frac{(2K+1)(K+1)}{4\pi K}}\frac{k^K}{(2K+1)!!}e\sqrt{\frac{4\pi}{2K+1}}r^K Y_K^q(\hat{\boldsymbol{r}}) \\
&= icA_0\sqrt{\frac{(2K+1)(K+1)}{4\pi K}}\frac{k^K}{(2K+1)!!}Q_{K,q}^{(\mathrm{el})},
\end{aligned}
\tag{2.28}
$$

with $c = \omega/k$ is the speed of light, and $Q_{K,q}^{(\mathrm{el})}$ in the last line is the electric multipole transition operator defined as

$$
Q_{K,q}^{(\mathrm{el})} = e\sqrt{\frac{4\pi}{2K+1}}r^K Y_K^q(\hat{\boldsymbol{r}}). \tag{2.29}
$$

This operator is a tensor operator with rank K.

The obtained electric multipole Hamiltonian $\mathcal{H}_{K,q}^{(\mathrm{el})}$ determines the transition rules of the electric multipole transitions. Using a parity operator $\mathcal{P}$, we find that

$$
\mathcal{P}\mathcal{H}_{K,q}^{(\mathrm{el})} = \mathcal{P}[r^K Y_K^q(\hat{\boldsymbol{r}})] = r^K Y_K^q(-\hat{\boldsymbol{r}}) = (-1)^K r^K Y_K^q(\hat{\boldsymbol{r}}), \tag{2.30}
$$

and thus the Hamiltonian has the parity of $(-1)^K$. Considering that the parity of the atomic state is determined by the angular momentum L as $(-1)^L$, and the parity must be conserved in the transition, the E1 transition ($K = 1$) is allowed only when $\Delta L = \pm 1$, and the E2 transition ($K = 2$) is allowed only when $\Delta L = 0, \ \pm 2$ and so on, where ΔL is the change of the angular momentum in the transition.

Additional transition rules arises from the coupling to the angular momentum of the light. The matrix element of the electric multipole transition operator $Q_{K,q}^{(\mathrm{el})}$ between the atomic states $|i\rangle = |\gamma, J, m\rangle$ and $|f\rangle = |\gamma', J', m'\rangle$ is obtained by employing the Wigner–Eckart theorem as

$$
\left\langle \gamma', J', m' \middle| Q_{K,q}^{(\mathrm{el})} \middle| \gamma, J, m \right\rangle = (-1)^{J'-m'}\begin{pmatrix} J' & K & J \\ -m' & q & m \end{pmatrix}\langle \gamma', J' || Q_{K,q}^{(\mathrm{el})} || \gamma, J\rangle. \tag{2.31}
$$

Here, γ is the radial wavefunction, J and m is the total angular momentum and its projection onto the quantization axis, respectively. The parentheses represent the Wigner 3j-symbol, and $\langle\gamma', J'||Q^{(\mathrm{el})}_{K,q}||\gamma, J\rangle$ is the reduced matrix element. The Wigner 3j-symbols follow the triangle condition $|J' - K| \leq J \leq J' + K$, which results in the additional selection rules. For example, the E1 transition ($K = 1$) prohibits $J = 0 \leftrightarrow J' = 0$, and the E2 transition ($K = 2$) prohibits $J = 0 \leftrightarrow J' = 0, 1$. We also find the restriction to K and q as $K \geq |\Delta J| = |J' - J|$ and $q = \Delta m = m' - m$.

2.1.2 Magnetic Multipole Transitions

The magnetic multipole component in the interaction Hamiltonian (Eq. (2.20)) is given by

$$\mathcal{H}^{(\mathrm{mg})}_{K,q} := \mathcal{H}^{(0)}_{K,q} = \frac{e}{m}\boldsymbol{p}\cdot\boldsymbol{a}^{(0)}_{K,q}(\hat{\boldsymbol{r}}) + \frac{e g_s}{2m}\boldsymbol{S}\cdot(\nabla\times\boldsymbol{a}^{(0)}_{K,q}(\hat{\boldsymbol{r}})). \tag{2.32}$$

Using the following relations

$$\boldsymbol{L} = \boldsymbol{r}\times\boldsymbol{p} = -i\hbar\boldsymbol{r}\times\nabla, \qquad \nabla = \hat{\boldsymbol{r}}\frac{\partial}{\partial r} - \frac{i}{\hbar r^2}\boldsymbol{r}\times\boldsymbol{L}, \tag{2.33}$$

the first term of the magnetic multipole Hamiltonian is written as

$$\begin{aligned}
\boldsymbol{p}\cdot\boldsymbol{a}^{(0)}_{K,q}(\hat{\boldsymbol{r}}) &= -i\hbar A_0\left(\hat{\boldsymbol{r}}\frac{\partial}{\partial r} - \frac{i}{\hbar r^2}\boldsymbol{r}\times\boldsymbol{L}\right)\cdot\left[j_K(kr)\boldsymbol{Y}^{(0)}_{K,q}(\hat{\boldsymbol{r}})\right] \\
&= -\hbar A_0\hat{\boldsymbol{r}}\cdot\frac{\boldsymbol{r}\times\nabla}{\sqrt{K(K+1)}}Y^q_K(\hat{\boldsymbol{r}})\frac{\partial j_K(kr)}{\partial r} - A_0\frac{j_K(kr)}{r^2}(\boldsymbol{r}\times\boldsymbol{L})\cdot\boldsymbol{Y}^{(0)}_{K,q}(\hat{\boldsymbol{r}}) \\
&= -iA_0\hat{\boldsymbol{r}}\cdot\frac{\boldsymbol{L}}{\sqrt{K(K+1)}}Y^q_K(\hat{\boldsymbol{r}})\frac{\partial j_K(kr)}{\partial r} - A_0\frac{j_K(kr)}{r^2}\left(\boldsymbol{r}\times\boldsymbol{Y}^{(0)}_{K,q}(\hat{\boldsymbol{r}})\right)\cdot\boldsymbol{L} \\
&= -A_0\frac{j_K(kr)}{r^2}\left(\boldsymbol{r}\times\boldsymbol{Y}^{(0)}_{K,q}(\hat{\boldsymbol{r}})\right)\cdot\boldsymbol{L}.
\end{aligned} \tag{2.34}$$

Here, we used the fact that $\boldsymbol{Y}^{(0)}_{k,q}(\hat{\boldsymbol{r}}) \propto \boldsymbol{L}$ and thus $(\boldsymbol{r}\times\boldsymbol{L})\cdot\boldsymbol{Y}^{(0)}_{K,q}(\hat{\boldsymbol{r}}) = \left(\boldsymbol{r}\times\boldsymbol{Y}^{(0)}_{K,q}(\hat{\boldsymbol{r}})\right)\cdot\boldsymbol{L}$ in the third line, and $\hat{\boldsymbol{r}}\cdot\boldsymbol{L} = 0$ in the last line. This expression is further simplified as

$$\begin{aligned}
\boldsymbol{p}\cdot\boldsymbol{a}^{(0)}_{K,q}(\hat{\boldsymbol{r}}) &= iA_0\frac{j_K(kr)}{r^2}\left[\boldsymbol{r}\times\frac{\boldsymbol{r}\times\nabla}{\sqrt{K(K+1)}}Y^q_K(\hat{\boldsymbol{r}})\right]\cdot\boldsymbol{L} \\
&= iA_0\frac{j_K(kr)}{r^2}\left[(\boldsymbol{r} - r^2\nabla)\frac{1}{\sqrt{K(K+1)}}Y^q_K(\hat{\boldsymbol{r}})\right]\cdot\boldsymbol{L} \\
&= iA_0\left[\frac{\partial J_K(kr)}{\partial r}\frac{\hat{\boldsymbol{r}}}{\sqrt{K(K+1)}}Y^q_K(\hat{\boldsymbol{r}}) - \frac{j_K(kr)}{\sqrt{K(K+1)}}\nabla Y^q_K(\hat{\boldsymbol{r}})\right]\cdot\boldsymbol{L} \\
&= i\frac{A_0}{\sqrt{K(K+1)}}\nabla[j_K(kr)Y^q_K(\hat{\boldsymbol{r}})]\cdot\boldsymbol{L}.
\end{aligned} \tag{2.35}$$

The second term in Eq. (2.32) is calculated as

$$\begin{aligned}\nabla \times \boldsymbol{a}^{(0)}_{K,q}(\hat{\boldsymbol{r}}) &= -iA_0(\nabla \times \boldsymbol{r} \times \nabla)j_K(kr)\frac{1}{\sqrt{K(K+1)}}Y^q_K(\hat{\boldsymbol{r}}) \\ &= -iA_0(\boldsymbol{r}\nabla^2 - \nabla(\boldsymbol{r}\cdot\nabla + 1))j_K(kr)\frac{1}{\sqrt{K(K+1)}}Y^q_K(\hat{\boldsymbol{r}}) \\ &= iA_0\sqrt{\frac{K+1}{K}}\nabla\left[j_K(kr)Y^q_K(\hat{\boldsymbol{r}})\right].\end{aligned} \tag{2.36}$$

In the above calculation, we used $\nabla^2 \boldsymbol{a}^{(0)}_{K,q} = 0$, and the first term in the second line vanishes. Finally, we obtain the magnetic multipole Hamiltonian as

$$\mathcal{H}^{\mathrm{mg}}_{K,q} = i\frac{eA_0}{m}\frac{1}{\sqrt{K(K+1)}}\nabla[j_K(kr)Y^q_K(\hat{\boldsymbol{r}})]\cdot\boldsymbol{L} + i\frac{eg_sA_0}{2m}\sqrt{\frac{K+1}{K}}\nabla[j_K(kr)Y^q_K(\hat{\boldsymbol{r}})]\cdot\boldsymbol{S}. \tag{2.37}$$

We again assume $kr \ll 1$ and use the asymptotic form of the spherical Bessel function to obtain the lowest order term of the magnetic multipole Hamiltonian as

$$\begin{aligned}\mathcal{H}^{\mathrm{mg}}_{K,q} &= iA_0\sqrt{\frac{(2K+1)(K+1)}{4\pi K}}\frac{k^K}{(2K+1)!!}\frac{e\hbar}{2m}\sqrt{\frac{4\pi}{2K+1}}\nabla\left[r^KY^q_K(\hat{\boldsymbol{r}})\right]\cdot\left[\frac{2}{K+1}\frac{\boldsymbol{L}}{\hbar} + g_s\frac{\boldsymbol{S}}{\hbar}\right] \\ &= iA_0\sqrt{\frac{(2K+1)(K+1)}{4\pi K}}\frac{k^K}{(2K+1)!!}Q^{(\mathrm{mg})}_{K,q},\end{aligned} \tag{2.38}$$

where $Q^{(\mathrm{mg})}_{K,q}$ is the magnetic multipole transition operator defined as

$$Q^{(\mathrm{mg})}_{K,q} = \frac{e\hbar}{2m}\sqrt{\frac{4\pi}{2K+1}}\nabla\left[r^KY^q_K(\hat{\boldsymbol{r}})\right]\cdot\left[\frac{2}{K+1}\frac{\boldsymbol{L}}{\hbar} + g_s\frac{\boldsymbol{S}}{\hbar}\right]. \tag{2.39}$$

This operator is also a tensor operator with rank K.

Similar to the electric multipole transitions, the magnetic multipole Hamiltonian $\mathcal{H}^{\mathrm{mg}}_{K,q}$ determines the transition rules of the magnetic multipole transitions. Using a parity operator $\mathcal{P}$, we find that

$$\mathcal{P}\mathcal{H}^{\mathrm{mg}}_{K,q} = \mathcal{P}\left[\nabla\left[r^KY^q_K(\hat{\boldsymbol{r}})\right]\cdot(\boldsymbol{L}+\boldsymbol{S})\right] = (-1)^{K+1}\nabla\left[r^KY^q_K(\hat{\boldsymbol{r}})\right]\cdot(\boldsymbol{L}+\boldsymbol{S}), \tag{2.40}$$

and the Hamiltonian has the parity of $(-1)^{K+1}$, which is opposite to the electric multipole Hamiltonian. Therefore, the M1 transition ($K = 1$) is allowed only when $\Delta L = 0$, and the M2 transition ($K = 2$) is allowed only when $\Delta L = \pm 1$ and so on.

The matrix element of the magnetic multipole transition operator $Q^{(\mathrm{mg})}_{K,q}$ between the atomic states $|i\rangle = |\gamma, J, m\rangle$ and $|f\rangle = |\gamma', J', m'\rangle$ is obtained by employing the Wigner–Eckart theorem as

$$\left\langle \gamma', J', m' \middle| Q^{(\mathrm{mg})}_{K,q} \middle| \gamma, J, m \right\rangle = (-1)^{J'-m'} \begin{pmatrix} J' & K & J \\ -m' & q & m \end{pmatrix} \langle \gamma', J' || Q^{(\mathrm{mg})}_{K,q} || \gamma, J \rangle . \tag{2.41}$$

This form is the same as the electric multipole transition matrix element in Eq. (2.31), and thus the selection rules on the total angular momentum of the magnetic multipole transitions are the same as those of the electric multipole transitions.

2.1.3 *Selection Rules and Relative Transition Strength of Multipole Transitions*

In the previous sections, we obtained the selection rules of the electric and magnetic multipole transitions. These are determined by Eqs. (2.30), (2.31), (2.40) and (2.41), and summarized in Table 2.1.

The relative transition strength between the electric and magnetic multipole transitions is estimated by comparing the Hamiltonians of Eqs. (2.28) and (2.38). The strength of each transitions is proportional to the square of the matrix element as

$$\frac{\Gamma^{(\mathrm{mg})}_{K,q}}{\Gamma^{(\mathrm{el})}_{K,q}} = \frac{\left|\mathcal{H}^{(\mathrm{mg})}_{K,q}\right|^2}{\left|\mathcal{H}^{(\mathrm{el})}_{K,q}\right|^2} \propto \left(\frac{\frac{e\hbar}{2m} r^{K-1}}{cer^K} \right)^2 = \left(\frac{\hbar}{2mcr} \right)^2 . \tag{2.42}$$

In atomic systems, the radius r is approximately expressed as a_0/Z_{eff}, where $a_0 = \hbar/(mc\alpha)$ is the Bohr radius, Z_{eff} is the effective charge of the nucleus, and α is the fine structure constant. Therefore, the relative strength of the magnetic multipole transitions to the electric multipole transitions is estimated as

$$\frac{\Gamma^{(\mathrm{mg})}_{K,q}}{\Gamma^{(\mathrm{el})}_{K,q}} \propto (\alpha Z_{\mathrm{eff}})^2 . \tag{2.43}$$

Since the fine structure constant $\alpha \simeq 1/137$ is a small number, the magnetic multipole transitions are much weaker than the electric multipole transitions. Furthermore, the relative strength of the electric multipole transitions with different K is estimated as

Table 2.1 Selection rules of multipole transitions

Transition	Parity	L	J	Example
E1	$\mathcal{P}_i = -\mathcal{P}_f$	$\Delta L = \pm 1$	$\Delta J = 0, \ \pm 1 \ (0 \nleftrightarrow 0)$	${}^1\mathrm{S}_0 \leftrightarrow {}^1\mathrm{P}_1$
M1	$\mathcal{P}_i = \mathcal{P}_f$	$\Delta L = 0$	$\Delta J = 0, \ \pm 1 \ (0 \nleftrightarrow 0)$	${}^3\mathrm{P}_1 \leftrightarrow {}^3\mathrm{P}_2$
E2	$\mathcal{P}_i = \mathcal{P}_f$	$\Delta L = 0, \ \pm 2$	$\Delta J = 0, \ \pm 1, \ \pm 2 \ (0 \nleftrightarrow 0, \ 1)$	${}^2\mathrm{S}_{1/2} \leftrightarrow {}^2\mathrm{D}_{5/2}$
M2	$\mathcal{P}_i = -\mathcal{P}_f$	$\Delta L = \pm 1$	$\Delta J = 0, \ \pm 1, \ \pm 2 \ (0 \nleftrightarrow 0, \ 1)$	${}^1\mathrm{S}_0 \leftrightarrow {}^3\mathrm{P}_2$

$$\frac{\Gamma^{(\mathrm{el})}_{K+1,q}}{\Gamma^{(\mathrm{el})}_{K,q}} = \frac{\left|\mathcal{H}^{(\mathrm{mg})}_{K,q}\right|^2}{\left|\mathcal{H}^{(\mathrm{el})}_{K,q}\right|^2} \propto (kr)^2. \tag{2.44}$$

The wavenumber $k = E/(\hbar c)$ is proportional to the transition energy E, which is approximately expressed as $E \propto Z_{\mathrm{eff}}^2 \alpha^2 mc^2$. Therefore, the relative strength of the electric multipole transitions with different K is estimated as

$$\frac{\Gamma^{(\mathrm{el})}_{K+1,q}}{\Gamma^{(\mathrm{el})}_{K,q}} \propto \left(Z_{\mathrm{eff}}^2 \alpha^2 \frac{mc}{\hbar} \cdot \frac{\hbar}{Z_{\mathrm{eff}} mc\alpha} \right)^2 = (\alpha Z_{\mathrm{eff}})^2 . \tag{2.45}$$

Equations (2.43) and (2.45) is the reason why the electric dipole transitions are the most dominant transitions in atomic systems.

2.1.4 Angular Dependence of Multipole Transition Strength

As we derived in the previous sections, the interaction Hamiltonian of the multipole transitions is written as

$$\mathcal{H}_{\mathrm{int}} = 4\pi \sum_{\mu=0,1} \sum_{K=1}^{\infty} \sum_{q=-K}^{K} i^{K-\mu} (\boldsymbol{Y}^{(\mu)}_{K,q}(\hat{\boldsymbol{k}}) \cdot \hat{\boldsymbol{\epsilon}}) \mathcal{H}^{(\mu)}_{K,q}. \tag{2.46}$$

In the above expression, we can find that the multipole transition also depends on the polarization and propagation axis of the light via the term of

$$C^{(\mu)}_{K,q} = \boldsymbol{Y}^{(\mu)}_{K,q}(\hat{\boldsymbol{k}}) \cdot \hat{\boldsymbol{\epsilon}}. \tag{2.47}$$

The transition strength of the multipole transitions is determined by $A^{(\mu)}_{K,q} = \left|C^{(\mu)}_{K,q}\right|^2$. To obtain explicit form of the angular dependence of the transition strength, we adopt following coordinates as

$$\begin{aligned} \hat{\boldsymbol{B}} &= \hat{\boldsymbol{z}} \\ \hat{\boldsymbol{k}} &= \sin(\theta)\hat{\boldsymbol{x}} + \cos(\theta)\hat{\boldsymbol{z}} \\ \hat{\boldsymbol{\epsilon}}_1 &= \hat{\boldsymbol{B}} \times \hat{\boldsymbol{k}} = \hat{\boldsymbol{y}} \\ \hat{\boldsymbol{\epsilon}}_2 &= \hat{\boldsymbol{k}} \times \hat{\boldsymbol{\epsilon}}_1 = -\cos(\theta)\hat{\boldsymbol{x}} + \sin(\theta)\hat{\boldsymbol{z}}, \end{aligned} \tag{2.48}$$

where $\hat{\boldsymbol{B}}$ is the quantization axis (Fig. 2.1a). The polarization of the light is represented using a normalized Jones vector (α, β) as $\hat{\boldsymbol{\epsilon}} = \alpha\hat{\boldsymbol{\epsilon}}_1 + \beta\hat{\boldsymbol{\epsilon}}_2$ with $|\alpha|^2 + |\beta|^2 = 1$. Note that in this coordinate the polarization and propagation axis of the light

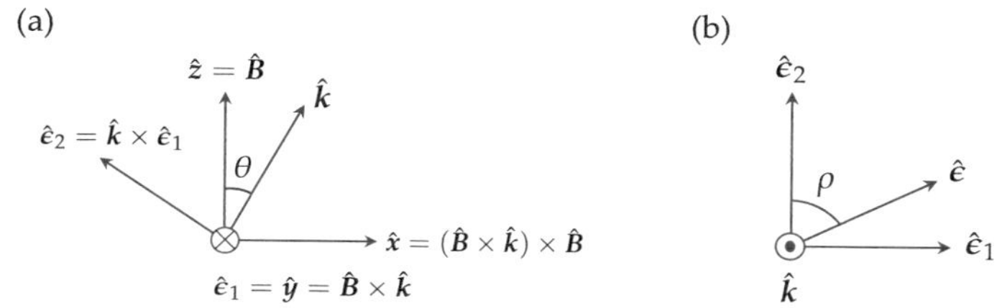

Fig. 2.1 Coordinates for angular dependency calculation of the multipole transition

are orthogonal to each other ($\hat{\boldsymbol{k}} \cdot \hat{\boldsymbol{\epsilon}} = 0$), and the azimuthal angle $\phi = 0$, that is $\boldsymbol{Y}^{(\mu)}_{K,q}(\hat{\boldsymbol{k}}) = \boldsymbol{Y}^{(\mu)}_{K,q}(\theta, 0)$. Recalling Eqs. (2.12) and (2.17) $\boldsymbol{Y}^{(\mu)}_{K,q}(\theta, 0)$ is written as

$$
\begin{aligned}
\boldsymbol{Y}^{(1)}_{K,q}(\theta, 0) &= \sum_{p=-1}^{1} \left[c^{(+)}_{K,q,p} Y^{q-p}_{K+1}(\theta, 0) + c^{(-)}_{K,q,p} Y^{q-p}_{K-1}(\theta, 0) \right] \hat{\boldsymbol{e}}_p, \\
\boldsymbol{Y}^{(0)}_{K,q}(\theta, 0) &= \sum_{p=-1}^{1} c^{(0)}_{K,q,p} Y^{q-p}_{K}(\theta, 0) \hat{\boldsymbol{e}}_p,
\end{aligned}
\tag{2.49}
$$

where the coefficients $c^{(\pm,0)}_{K,q,p}$ is the weight of the projection of the spherical harmonics onto the p direction and can be obtained from the definition of the vector spherical harmonics (Eq. (2.12)) as

$$
\begin{aligned}
c^{(+)}_{K,q,p} &= (-1)^{K-q} \sqrt{K} \begin{pmatrix} K & 1 & K+1 \\ -q & p & q-p \end{pmatrix} \\
c^{(-)}_{K,q,p} &= (-1)^{K-q} \sqrt{K+1} \begin{pmatrix} K & 1 & K-1 \\ -q & p & q-p \end{pmatrix} \\
c^{(0)}_{K,q,p} &= (-1)^{K-q} \sqrt{2K+1} \begin{pmatrix} K & 1 & K \\ -q & p & q-p \end{pmatrix}.
\end{aligned}
\tag{2.50}
$$

The unit vectors $\hat{\boldsymbol{e}}_\pm$ and $\hat{\boldsymbol{e}}_0$ correspond to $\sigma^\pm$- and π-transitions, respectively. Equation (2.49) means that in the multipole transitions the light can provide the angular momentum of $m_s = 0, \ \pm 1$ by the photon spin, which is represented by $\hat{\boldsymbol{e}}_p$, and additional angular momentum of $m_l = q, \ q \pm 1$, which is represented by the spherical harmonics. Considering the angular momentum conservation, the transition must satisfy $\Delta m = m_s + m_l$, where Δm is the difference between the projection of the angular momentum of the initial and final states.

In the following, we visualize the angular dependence of the transition strength based on Eqs. (2.47) and (2.49). For simplicity, we consider linearly polarized light with a polarization vector of

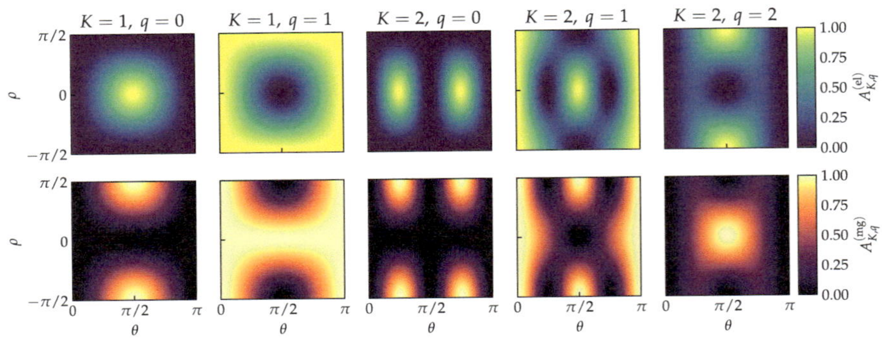

Fig. 2.2 Angular dependence of the multipole transition strength calculated from Eqs. (2.47) and (2.49). The top and bottom row represents electric and magnetic multipole transition, respectively

$$\begin{aligned}\hat{\boldsymbol{\epsilon}}_{\text{lin}} &= \sin(\rho)\hat{\boldsymbol{\epsilon}}_1 + \cos(\rho)\hat{\boldsymbol{\epsilon}}_2 \\ &= \sin(\rho)\hat{\boldsymbol{y}} + \cos(\rho)\left(-\cos(\theta)\hat{\boldsymbol{x}} + \sin(\theta)\hat{\boldsymbol{z}}\right) \\ &= \frac{i\sin(\rho) + \cos(\rho)\cos(\theta)}{\sqrt{2}}\hat{\boldsymbol{e}}_1 + \cos(\rho)\sin(\theta)\hat{\boldsymbol{e}}_0 + \frac{i\sin(\rho) - \cos(\rho)\cos(\theta)}{\sqrt{2}}\hat{\boldsymbol{e}}_{-1},\end{aligned} \tag{2.51}$$

where ρ is the angle between the polarization vector and its projection onto the x-z plane (Fig. 2.1b). Figure 2.2 shows the angular dependence of the transition strength for the E1, E2, M1, and M2 transitions. We can find that the transition strength of the electric and magnetic multipole transitions of the same order has the same angular dependence, up to shift of the angle ρ by $\pi/2$. This is because the electric and magnetic fields are orthogonal to each other.

2.2 Atomic Polarizability and Dipole Potential

As we discussed in Sect. 2.1.3, the atom-light interaction is approximated by the electric dipole transitions in most cases. The induced electric dipole $\boldsymbol{d}$ by the oscillating electic field $\boldsymbol{E}(\boldsymbol{r}, t) = \hat{\boldsymbol{\epsilon}} E(\boldsymbol{r}) \exp(-i\omega t) + \text{c.c}$ is given by

$$\boldsymbol{d}(\boldsymbol{r}, t) = \alpha(\omega)\boldsymbol{E}(\boldsymbol{r}, t). \tag{2.52}$$

Here, $\alpha(\omega)$ is the atomic polarizability, which depends on the frequency of the light. As we will see later, the real and imaginary part of the polarizability is related to the dipole potential and the scattering rate of the off-resonant light, respectively.

In this section, we will calculate the atomic polarizability in a two-level system and introduce optical dipole traps, which are essential in this study. After that we will extend it to a multi-level system.

2.2.1 Two-Level System

The polarizability in a two-level system can be understood by the Lorentz-oscillator model, where the electron is bound to the core with an oscillation frequency ω_0 and interacts with the light field with a frequency ω. This classical model gives the polarizability as [5]

$$\alpha(\omega) = 6\pi\epsilon_0 c^3 \frac{\Gamma}{\omega_0^2\left(\omega_0^2 - \omega^2 - i\Gamma\frac{\omega^3}{\omega_0^2}\right)}, \tag{2.53}$$

where Γ is the decay rate of the excited state.

The induced dipole potential (also referred as light shift) U_{dip} can be calculated as

$$U_{\text{dip}}(\boldsymbol{r}) = -\frac{1}{2}\frac{1}{T}\int_0^T \boldsymbol{d}(\boldsymbol{r},t)\cdot\boldsymbol{E}(\boldsymbol{r},t)\mathrm{d}t = -\frac{1}{2}\mathrm{Re}\left[\alpha(\omega)\right]|\boldsymbol{E}(\boldsymbol{r})|^2 = -\frac{1}{2\epsilon_0 c}\mathrm{Re}\left[\alpha(\omega)\right]I(\boldsymbol{r}), \tag{2.54}$$

where ϵ_0 is the vacuum permittivity, c is the speed of light, and $I(\boldsymbol{r})$ is the intensity of the light. The factor of $1/2$ means that the dipole is induced by the electric field. The corresponding dipole force is then calculated as

$$\boldsymbol{F}(\boldsymbol{r}) = -\nabla U_{\text{dip}}(\boldsymbol{r}) = \frac{1}{2\epsilon_0 c}\nabla\mathrm{Re}\left[\alpha(\omega)\right]I(\boldsymbol{r}). \tag{2.55}$$

If the polarizability is positive, the force is attractive, and if the polarizability is negative, the force is repulsive. The associated scattering rate γ_{sc}, which corresponds to the absorbed power per photon, is calculated as

$$\gamma_{\text{sc}} = \frac{1}{\hbar\omega}\frac{1}{T}\int_0^T\left[\frac{\partial}{\partial t}\boldsymbol{d}(\boldsymbol{r},t)\right]\cdot\boldsymbol{E}(\boldsymbol{r},t)\mathrm{d}t = \frac{1}{\hbar\epsilon_0 c}\mathrm{Im}\left[\alpha(\omega)\right]I(\boldsymbol{r}). \tag{2.56}$$

By substituting Eq. (2.53) into Eqs. (2.54) and (2.56), we can obtain the dipole potential and the scattering rate in a two-level system as

$$U_{\text{dip}}(\boldsymbol{r}) = -\frac{3\pi c^2\Gamma}{2\omega_0^2}\frac{\omega_0^2-\omega^2}{(\omega_0^2-\omega^2)^2-\Gamma^2\frac{\omega^6}{\omega_0^4}}I(\boldsymbol{r}) \simeq -\frac{3\pi c^2}{2\omega_0^3}\left(\frac{\Gamma}{\omega_0-\omega}+\frac{\Gamma}{\omega_0+\omega}\right)I(\boldsymbol{r}), \tag{2.57}$$

$$\gamma_{\text{sc}} = \frac{6\pi c^2\Gamma^2}{\hbar\omega_0^2}\frac{\frac{\omega^3}{\omega_0^2}}{\left(\omega_0^2-\omega^2\right)^2-\Gamma^2\frac{\omega^6}{\omega_0^4}}I(\boldsymbol{r}) \simeq \frac{3\pi c^2}{2\hbar\omega_0^3}\left(\frac{\omega}{\omega_0}\right)^3\left(\frac{\Gamma}{\omega_0-\omega}+\frac{\Gamma}{\omega_0-\omega}\right)^2 I(\boldsymbol{r}). \tag{2.58}$$

In the last step, we neglected the term of $\Gamma^2\omega^6/\omega_0^4$ in the denominator, which is valid for far off-resonant light. We introduce the detuning $\Delta = \omega - \omega_0$ and apply the rotating wave approximation ($|\omega_0 - \omega| \ll \omega_0 + \omega$, and $\omega \sim \omega_0$) to obtain simpler expressions as

$$U_{\text{dip}}(\boldsymbol{r}) = -\frac{3\pi c^2}{2\omega_0^3}\frac{\Gamma}{\Delta}I(\boldsymbol{r}), \tag{2.59}$$

$$\gamma_{\text{sc}} = \frac{3\pi c^2}{2\hbar\omega_0^3}\left(\frac{\Gamma}{\Delta}\right)^2 I(\boldsymbol{r}). \tag{2.60}$$

We can see that the dipole potential is attractive for red-detuned light ($\Delta > 0$), which is used for optical trapping of the atoms. A gaussian beam is usually used for this purpose (optical dipole trap), and the intensity of the gaussian beam is given by

$$I(\rho, z) = I_0\left(\frac{w_0^2}{w(z)^2}\right)\exp\left(-\frac{2\rho^2}{w(z)^2}\right) \tag{2.61}$$

where ρ is the distance from the beam center, z is the position along the beam propagation axis, w_0 is the beam waist at the focus ($z = 0$), $w(z) = w_0\sqrt{1 + (z/z_R)^2}$ is the beam waist at the position z, and $z_R = \pi w_0^2/\lambda$ is the Rayleigh length. I_0 is the peak intensity of the beam defined as $I_0 = 2P/(\pi w_0^2)$ with beam power P. By substituting Eq. (2.61) into Eq. (2.59) and using Taylor expansion around $z = 0$, we can obtain the dipole potential as a harmonic trap as

$$U_{\text{dip}}(\rho, z) = -U_0\left[1 - 2\left(\frac{\rho}{w_0}\right)^2 - \left(\frac{z}{z_R}\right)^2\right] = -U_0 + \frac{1}{2}m\omega_\rho^2 r^2 + \frac{1}{2}m\omega_z z^2, \tag{2.62}$$

where $U_0 = 3\pi c^2\Gamma/(2\omega_0^3\Delta)I_0$ is the potential depth, m is the atomic mass, and ω_ρ and ω_z are the trap frequencies along the radial and axial direction of the trap, which are defined as

$$\omega_\rho := \sqrt{\frac{4U_0}{mw_0^2}}, \quad \omega_z := \sqrt{\frac{2U_0}{mz_R^2}}. \tag{2.63}$$

Figure 2.3 shows the spatial distribution of the dipole potential made by a gaussian beam. The potential is harmonic along the radial axial directions.

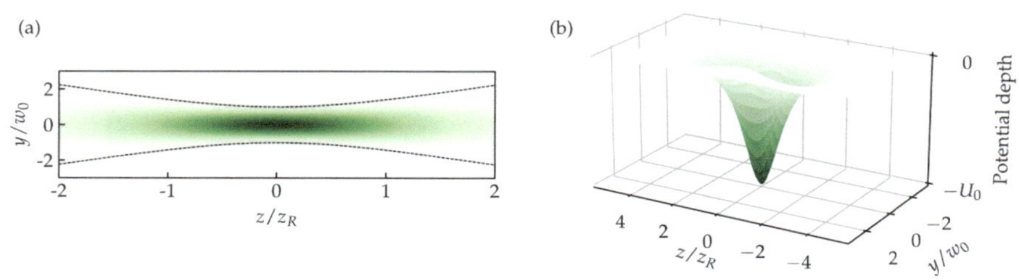

Fig. 2.3 Dipole potential of a gaussian beam. **a** Two-dimensional plot. The dashed lines represent $1/e^2$ radius of the trapping beam. **b** Three-dimensional plot

2.2.2 *Multi-level System*

In reality, the atom has multiple energy levels, and thus we must take into account these levels to calculate the polarizability. Hereafter, we refer to the polarizability as the real part of the complex polarizability. The atomic polarizability is rank 2 tensor arising from the ac Stark shift Hamiltonian, and can be decomposed into the scalar, vector, and tensor parts, which are given by [6]

$$\alpha^S_{nJ} = \frac{1}{\sqrt{3(2J+1)}}\alpha^{(0)}_{nJ}, \tag{2.64}$$

$$\alpha^V_{nJ} = -\sqrt{\frac{2J}{(J+1)(2J+1)}}\alpha^{(1)}_{nJ}, \tag{2.65}$$

$$\alpha^T_{nJ} = -\sqrt{\frac{2J(2J-1)}{3(J+1)(2J+1)(2J+3)}}\alpha^{(2)}_{nJ}, \tag{2.66}$$

with

$$\alpha^{(K)}_{nJ} = (-1)^{K+J+1}\sqrt{2K+1}\sum_{n',J'}(-1)^{J'}\begin{Bmatrix}1 & K & 1\\ J & J' & J\end{Bmatrix}\left|\langle n'J'||\boldsymbol{d}||nJ\rangle\right|^2 \times \frac{2\pi}{\epsilon_0 ch^2}\mathrm{Re}\left[\frac{1}{\omega_{n'J'nJ}-\omega-i\gamma_{n'J'nJ}/2}+\frac{(-1)^K}{\omega_{n'J'nJ}+\omega-i\gamma_{n'J'nJ}/2}\right]. \tag{2.67}$$

Here, n, J is the principal quantum number and total angular momentum of the atomic state $|n, J\rangle$, $\omega_{n'J'nJ}$ is the transition frequency between the states $|n', J'\rangle$ and $|n, J\rangle$, $\gamma_{n'J'nJ}$ is the decay rate of the state $|n', J'\rangle$, $\{\cdot\}$ is the Wigner 6j-symbol, and $\langle n'J'||\boldsymbol{d}||nJ\rangle$ is the reduced dipole matrix element. With these polarizabilities, the dipole potential operator is written as

$$\hat{\mathcal{U}}_{\mathrm{dip}} = -\frac{h}{4}\left[\alpha^S_{nJ} - i\alpha^V_{nJ}\frac{(\boldsymbol{\epsilon}^*\times\boldsymbol{\epsilon})\cdot\boldsymbol{J}}{2J} + \alpha^T_{nJ}\frac{3[(\boldsymbol{\epsilon}^*\cdot\boldsymbol{J})(\boldsymbol{\epsilon}\cdot\boldsymbol{J})+(\boldsymbol{\epsilon}^*\cdot\boldsymbol{J})(\boldsymbol{\epsilon}\cdot\boldsymbol{J})]-2\boldsymbol{J}^2}{2J(2J-1)}\right]I(\boldsymbol{r}), \tag{2.68}$$

and the dipole potential is calculated as $U_{\mathrm{dip}}(\boldsymbol{r}) = \langle n, J|\hat{\mathcal{U}}_{\mathrm{dip}}|n, J\rangle$. Note that we conventionally chose the coefficient of the dipole potential as $h/4$.

If the atomic levels have hyperfine structures $\boldsymbol{F} = \boldsymbol{J} + \boldsymbol{I}$, where $\boldsymbol{I}$ is the nuclear spin, the polarizability and dipole potential are written as

$$\alpha^S_{nJF} = \alpha^S_{nJ} = \frac{1}{\sqrt{3(2J+1)}}\alpha^{(0)}_{nJ}, \tag{2.69}$$

$$\alpha^V_{nJF} = (-1)^{J+I+F}\sqrt{\frac{2F(2F+1)}{F+1}}\begin{Bmatrix}F & 1 & F\\ J & I & J\end{Bmatrix}\alpha^{(1)}_{nJ}, \tag{2.70}$$

$$\alpha^T_{nJF} = -(-1)^{J+I+F}\sqrt{\frac{2F(2F-1)(2F+1)}{3(F+1)(2F+3)}}\begin{Bmatrix} F & 2 & F \\ J & I & J \end{Bmatrix}\alpha^{(2)}_{nJ}, \tag{2.71}$$

and

$$\hat{\mathcal{U}}_{\text{dip}} = -\frac{h}{4}\left[\alpha^S_{nJF} - i\alpha^V_{nJF}\frac{(\boldsymbol{\epsilon}^* \times \boldsymbol{\epsilon})\cdot \boldsymbol{F}}{2F} + \alpha^T_{nJF}\frac{3[(\boldsymbol{\epsilon}^*\cdot\boldsymbol{F})(\boldsymbol{\epsilon}\cdot\boldsymbol{F}) + (\boldsymbol{\epsilon}^*\cdot\boldsymbol{F})(\boldsymbol{\epsilon}\cdot\boldsymbol{F})] - 2F^2}{2F(2F-1)}\right] I(\boldsymbol{r}). \tag{2.72}$$

If the off-diagonal coupling is much smaller than the Zeeman coupling, and the energy levels are well described as the Zeeman sublevels $|F, m_F\rangle$, the dipole potential is further simplified as

$$\begin{aligned} U_{\text{dip}}(\boldsymbol{r}) &= -\frac{h}{4}\left[\alpha^S_{nJF} - 2\text{Im}(\epsilon^*_x\epsilon_y)\alpha^V_{nJF}\frac{m_F}{2F} - (1 - 3|\epsilon_z|^2)\alpha^T_{nJF}\frac{3m_F^2 - F(F+1)}{2F(2F-1)}\right] I(\boldsymbol{r}) \\ &= -\frac{h}{4}\left[\alpha^S_{nJF} - \sin(2\gamma)\alpha^V_{nJF}\frac{m_F}{2F} - (1 - 3\cos^2\theta)\alpha^T_{nJF}\frac{3m_F^2 - F(F+1)}{2F(2F-1)}\right] I(\boldsymbol{r}), \end{aligned} \tag{2.73}$$

where θ is the angle between the light polarization axis and the quantization axis, and γ is the ellipticity angle of the light, which represents the elliptical polarization as

$$\boldsymbol{\epsilon} = \cos(\gamma)\hat{\boldsymbol{x}} + i\sin(\gamma)\hat{\boldsymbol{y}}. \tag{2.74}$$

As seen in the above expression, while the scalar term is isotropic, the vector and tensor terms depend on the polarization of the light. The vector term is vanished when the light is linearly polarized ($\gamma = 0$), and is maximized when the light is circulary polarized ($\gamma = \pi/4$). The tensor term also has the tunability of the light shift via the angle between the polarization and quantization axis.

These properties act as "tuning knobs" of the dipole potential. One can utilize these to make a magic condition for a certain atomic transition, where total differential light shift between the transition states is canceled out. An elliptic polarization, which makes a magic condition, is called magic polarization [7–9]. Similarly, if the magic condition is satisfied by adjusting the angle between the polarization and quantization axis, the angle is called magic angle [10–14]. These magic conditions are useful for high-fidelity excitation, cooling, and imaging of the atoms.

2.3 Coherent Dynamics in Atom-Light Interaction

Coherent manipulation of the atomic internal state by light is essential in quantum information processing in neutral atom systems. In this section, we will discuss the coherent dynamics of the atom-light interaction. For simplicity, we consider a two-level system consisting of the ground state $|g\rangle$ and the excited state $|e\rangle$ with a energy difference of $\hbar\omega_0$. The total Hamiltonian of the system is given by

$$\begin{aligned}\mathcal{H} &= \hbar\omega_0 \,|e\rangle\langle e| + \langle g|\,\mathcal{H}_{\text{int}}\,|e\rangle \,|g\rangle\langle e| + \langle e|\,\mathcal{H}_{\text{int}}\,|g\rangle\,|e\rangle\langle g| \\ &= \hbar\omega_0\,|e\rangle\langle e| + \frac{\hbar\Omega}{2}\left(e^{-i\omega t}\,|e\rangle\langle g| + e^{i\omega t}\,|g\rangle\langle e|\right),\end{aligned} \tag{2.75}$$

where $\mathcal{H}_{\text{int}}$ is given as Eq. (2.4), Ω is the Rabi frequency, and ω is the frequency of the light. In the rotating frame with an angular frequency ω_0, the Hamiltonian is given by the transformation with a unitary operator $U = \exp(i\omega t\,|e\rangle\langle e|)$ as

$$\tilde{\mathcal{H}} = U\mathcal{H}U^\dagger + i\hbar(\partial_t U)U^\dagger = -\hbar\Delta\,|e\rangle\langle e| + \frac{\hbar\Omega}{2}\left(|e\rangle\langle g| + |g\rangle\langle e|\right). \tag{2.76}$$

Here, $\Delta = \omega_0 - \omega$ is the detuning of the light.

The time-evolution of the system $\rho(t) = \rho_{gg}\,|g\rangle\langle g| + \rho_{ee}\,|e\rangle\langle e| + \rho_{ge}\,|g\rangle\langle e| + \rho_{eg}\,|e\rangle\langle g|$ is described by the master equation as

$$\frac{\partial\rho}{\partial t} = -\frac{i}{\hbar}[\tilde{\mathcal{H}}, \rho] + \Gamma\mathcal{D}[|g\rangle\langle e|]\,\rho, \tag{2.77}$$

where Γ is the decay rate of the excited state, and $\mathcal{D}[c]\rho$ is the Lindblad superoperator defined as

$$\mathcal{D}[c]\rho = c\rho c^\dagger - \frac{1}{2}\left(c^\dagger c\rho + \rho c^\dagger c\right). \tag{2.78}$$

The time evolution of each component is then written as

$$\begin{aligned}\partial_t\rho_{gg} &= -i\frac{\Omega}{2}(\rho_{eg} - \rho_{ge}) + \Gamma\rho_{ee}, \\ \partial_t\rho_{ee} &= i\frac{\Omega}{2}(\rho_{eg} - \rho_{ge}) - \Gamma\rho ee, \\ \partial_t\rho_{ge} &= -\left(\frac{\Gamma}{2} + i\Delta)\rho_{ge} - i\frac{\Omega}{2}(\rho_{ee} - \rho_{gg}\right), \\ \partial_t\rho_{eg} &= -\left(\frac{\Gamma}{2} - i\Delta)\rho_{ge} + i\frac{\Omega}{2}(\rho_{ee} - \rho_{gg}\right).\end{aligned} \tag{2.79}$$

These equations are called optical Bloch equations. When the atom is in the ground state, and the light is on resonance of the transition, the dynamics of $\langle\sigma_z\rangle = \rho_{ee} - \rho_{gg}$ is written as [4]

$$\langle\sigma_z(t)\rangle = -1 + \frac{\Omega^2}{\Omega^2 + \Gamma^2/2}\left[1 - e^{-(3\Gamma/4)t}\left(\cos(\Omega_\Gamma t) + \frac{3\Gamma}{4\Omega_\Gamma}\sin(\Omega_\Gamma t)\right)\right], \tag{2.80}$$

with $\Omega_\Gamma = \sqrt{\Omega^2 - (\Gamma/4)^2}$ as shown in Fig. 2.4. The steady-state solution is given by

$$\rho_{ee}(t\to\infty) = 1 - \rho_{gg}(t\to\infty) = \frac{\frac{\Omega^2}{\Gamma^2}}{1 + \left(\frac{2\Delta}{\Gamma}\right)^2 + \frac{2\Omega^2}{\Gamma^2}} = \frac{1}{2}\frac{I/I_{\text{sat}}}{1 + (2\Delta/\Gamma)^2 + I/I_{\text{sat}}}. \tag{2.81}$$

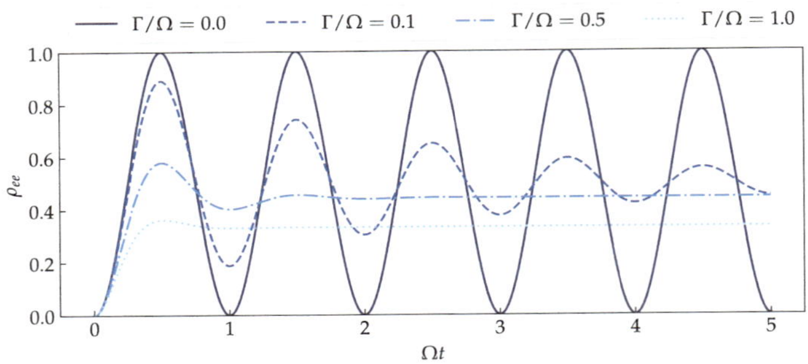

Fig. 2.4 Rabi oscillations for various Rabi frequencies with zero detuning

Here, we introduced the saturation intensity $I_{\mathrm{sat}} = \pi hc\Gamma/(3\lambda^3)$, where λ is the wavelength of the light, and used the relation $I/I_{\mathrm{sat}} = 2\Omega^2/\Gamma^2$.

2.4 Center-of-Mass Motion of Atoms in Optical Transitions

In addition to the coherent dynamics of the atom-light interaction, the mechanical effects of the light on the atom are also pivotal in cooling and trapping of the atoms.

2.4.1 *Atoms in Free Space*

First, we consider the center-of-mass motion of the atom moving in one-dimensional free space with a velocity v and momentum $p = mv$. The scattering rate during irradiation of near-resonant light with a detuning Δ and an intensity I is given by

$$R_{\mathrm{sc}}(v) = \Gamma\rho_{ee}(t \to \infty) = \frac{\Gamma}{2}\frac{I/I_{\mathrm{sat}}}{1 + (2(\Delta - kv)/\Gamma)^2 + I/I_{\mathrm{sat}}}. \tag{2.82}$$

Here, we take into account the Doppler shift of $-kv$ in the detuning. The corresponding radiation force acting on the atom is given by

$$F_{\mathrm{rad}} = \hbar k R_{\mathrm{sc}}. \tag{2.83}$$

When two counter-propagating beams are irradiated, the energy change of the atom $(\mathrm{d}E)_{\mathrm{cool}}$ in time $\mathrm{d}t$, which is induced by the light irradiation, is given by

$$(\mathrm{d}E)_{\mathrm{cool}} = F_{\mathrm{rad}} v \mathrm{d}t = \hbar k \left(R_{\mathrm{sc}}(v) - R_{\mathrm{sc}}(-v)\right) \mathrm{d}t. \tag{2.84}$$

When the Doppler shift is much smaller than the detuning, the scattering rate is approximated as

$$R_{sc}(v) = R_{sc}(0) + v\frac{dR_{sc}}{dv}|_{v=0} = R_{sc}(0) + kv\frac{4(\Delta/\Gamma)I/I_{sat}}{\left(1 + (2\Delta/\Gamma)^2 + I/I_{sat}\right)^2} \tag{2.85}$$

Substituting Eq. (2.85) into Eq. (2.87), we obtain

$$(\mathrm{d}E)_{cool} = \frac{32(\Delta/\Gamma)I/I_{sat}}{(1 + (2\Delta/\Gamma)^2 + I/I_{sat})^2}\frac{E_R}{\hbar}E\mathrm{d}t, \tag{2.86}$$

where $E_R = \hbar^2k^2/(2m)$ is the recoil energy. When the beams are red-detuned ($\Delta < 0$), the energy change is negative, meaning that the beams cool the atom. On the other hand, the atom absorbs and emits a photon with a momentum $\hbar k$ randomly, resulting in a random walk in momentum space and energy increase of the atom $(\mathrm{d}E)_{heat}$ as

$$(\mathrm{d}E)_{heat} = 2E_R\left(R_{sc}(v) + R_{sc}(-v)\right)\mathrm{d}t = 4E_R\frac{\Gamma}{2}\frac{I/I_{sat}}{1 + (2\Delta/\Gamma)^2 + I/I_{sat}}\mathrm{d}t. \tag{2.87}$$

The factor of 2 arises from the random process of the absorption and emission of photons. We finally obtain the net energy change of the atom from Eqs. (2.84) and (2.87) as

$$\mathrm{d}Et = \frac{(\mathrm{d}E)_{heat} + (\mathrm{d}E)_{cool}}{\mathrm{d}t} = 2E_R\frac{I/I_{sat}}{1 + (2\Delta/\Gamma)^2 + I/I_{sat}}\left(\Gamma + \frac{16(\Delta/\Gamma)I/I_{sat}}{1 + (2\Delta/\Gamma)^2 + I/I_{sat}}\frac{E}{\hbar}\right). \tag{2.88}$$

In the steady state and low intensity limit ($I/I_{sat} \to 0$), the energy becomes

$$E = -\frac{\hbar\Gamma}{16}\frac{1 + (2\Delta/\Gamma)^2}{\Delta/\Gamma}, \tag{2.89}$$

which has a minimum at $E_{min} = \hbar\Gamma/4$ when $\Delta = -\Gamma/2$. The corresponding temperature is $T_D = 2E/k_B = \hbar\Gamma/(2k_B)$, where k_B is the Boltzmann constant. This temperature T_D is called the Doppler limit, which is the lowest temperature that can be achieved by the Doppler cooling.

2.4.2 Atoms in a Harmonic Trap

Next, we consider the center-of-mass motion of the atom in a harmonic trap with a trap frequency of ω_ρ. In a harmonic trap, the motion of the atom is quantized according to following Hamiltonian as

$$\mathcal{H}_{trap} = \frac{\hat{p}^2}{2m} + \frac{1}{2}m\omega_\rho^2\hat{x}^2 = \hbar\omega_\rho\left(\hat{a}^\dagger\hat{a} + \frac{1}{2}\right), \tag{2.90}$$

where $\hat{x},\ \hat{p}$ is the position and momentum operator, and related to the annihilation and creation operators as

$$\begin{aligned}\hat{x} &= \sqrt{\frac{\hbar}{2m\omega_\rho}}(\hat{a}+\hat{a}^\dagger) = x_0(\hat{a}+\hat{a}^\dagger)\\ \hat{p} &= i\sqrt{\frac{\hbar m\omega_\rho}{2}}(\hat{a}^\dagger-\hat{a}) = p_0(\hat{a}^\dagger-\hat{a}).\end{aligned} \tag{2.91}$$

Assuming that the atom has only two internal states $|g\rangle$ and $|e\rangle$, the atomic states are labeled as $|g, n_g\rangle$ and $|e, n_e\rangle$, where $n_g,\ n_e$ is the vibrational quantum number for each internal state. Then the transition matrix element can be written as

$$\langle g, n_g|\mathcal{H}_{\rm int}|e, n_e\rangle = \hbar\Omega_0\left\langle n_g\left|e^{ik\hat{x}}\right|n_e\right\rangle = \hbar\Omega_0\left\langle n_g\left|e^{i\eta(\hat{a}+\hat{a}^\dagger)}\right|n_e\right\rangle, \tag{2.92}$$

where Ω_0 is the Rabi frequency for the carrier transition ($n_g = n_e$), and $\eta = kx_0 = \sqrt{E_R/(\hbar\omega_\rho)}$ is the Lamb-Dicke factor. When the trap frequency is much larger than the recoil frequency ($\omega_\rho \gg E_R/\hbar$), which means that the atoms are tightly confined in the trap, the Lamb-Dicke factor is much smaller than 1, and the matrix element is approximated as

$$\Omega_0\left\langle n_g\left|e^{i\eta(\hat{a}+\hat{a}^\dagger)}\right|n_e\right\rangle \simeq \Omega_0\left\langle n_g\right|(1+i\eta(\hat{a}+\hat{a}^\dagger))\left|n_e\right\rangle = \begin{cases}\Omega_0 & (n_g = n_e)\\ i\eta\sqrt{n_e}\Omega_0 & (n_e = n_g+1)\\ i\eta\sqrt{n_g}\Omega_0 & (n_e = n_g-1)\\ 0 & (\text{otherwise})\end{cases} \tag{2.93}$$

The expressions means that the transition with a motional state change, which are called sidebands, is suppressed by the factor $\eta \ll 1$. This regime is called the Lamb-Dicke regime. If we calculate the matrix element without the approximation, we obtain [15]

$$\Omega_0\left\langle n_g\left|e^{i\eta(\hat{a}+\hat{a}^\dagger)}\right|n_e\right\rangle = \Omega_0 e^{-\eta^2/2}\eta^{\Delta n}\sqrt{\frac{n_<!}{n_>!}}L_{n_<}^{\Delta n}(\eta^2), \tag{2.94}$$

where $n_< = \min(n_g, n_e),\ n_> = \max(n_g, n_e),\ \Delta n = |n_g - n_e|$, and $L_{n_<}^{\Delta n}(\eta^2)$ is the Laguerre polynomial as

$$L_n^\alpha(x) = \sum_{k=0}^{n}\binom{n+\alpha}{n-k}\frac{(-x)^k}{k!}. \tag{2.95}$$

It is important to note that the above calculation assumes the trap satisfies the magic condition, meaning the trap frequency is the same for each internal state. Under this condition, the wavefunctions of motional states with the same vibrational quantum number in different internal states are identical. Specifically, this implies that

$\langle n_g | n_e \rangle = 1$ when $n_g = n_e$. However, in reality, this assumption does not always hold, and the wavefunctions of the motional states differ between internal states. In such cases, $\langle n_g | n_e \rangle \neq 1$ for $n_g = n_e$ and $\langle n_g | n_e \rangle \neq 0$ for $n_g \neq n_e$, explicitly written as [16]

$$\langle 0_g | 0_e \rangle = \left[\sum_{m=0}^{\infty} \frac{(2m-1)!!}{(2m)!!} \left(\frac{\omega_e - \omega_g}{\omega_e + \omega_g} \right)^{2m} \right]^{-1/2} \tag{2.96}$$

$$\langle n_g | 0_e \rangle = (-1)^{n_g/2} \sqrt{\frac{(n_g-1)!!}{n_g!!}} \left(\frac{\omega_e - \omega_g}{\omega_e + \omega_g} \right)^{n_g/2} \langle 0_g | 0_e \rangle \quad (\text{for even } n_g) \tag{2.97}$$

$$\langle n_g | 0_e \rangle = 0 \quad (\text{for odd } n_g) \tag{2.98}$$

$$\langle n_g | n_e \rangle = \frac{1}{2\sqrt{n_e}} \left[\alpha_+ \sqrt{n_g} \langle n_g - 1 | n_e - 1 \rangle + \alpha_- \sqrt{n_g + 1} \langle n_g + 1 | n_e - 1 \rangle \right], \tag{2.99}$$

where ω_g, ω_e is the trap frequency for the ground and excited states, respectively, and

$$\alpha_{\pm} = \sqrt{\frac{\omega_e}{\omega_g}} \pm \sqrt{\frac{\omega_g}{\omega_e}}. \tag{2.100}$$

This results in unintended spreading of the vibrational quantum number during the dynamics when the sidebands are not well-resolved for the transition. This leads to the heating of the atoms in the trap.

References

1. W.R. Johnson, *Atomic Structure Theory: Lectures on Atomic Physics* (Springer, 2007). https://doi.org/10.1007/978-3-540-68013-0. https://link.springer.com/book/10.1007/978-3-540-68013-0
2. J. Trautmann, The magnetic quadrupole transition in neutral strontium. Ph.D. thesis, Ludwig-Maximilians-Universität München (2022). https://edoc.ub.uni-muenchen.de/30737/
3. J. Trautmann, D. Yankelev, V. Klüsener, A.J. Park, I. Bloch, S. Blatt, Phys. Rev. Res. **5**(1), 013219 (2023). https://doi.org/10.1103/PhysRevResearch.5.013219. https://link.aps.org/doi/10.1103/PhysRevResearch.5.013219
4. D.A. Steck, *Quantum and Atom Optics* (2007). https://atomoptics.uoregon.edu/~dsteck/teaching/quantum-optics/
5. R. Grimm, M. Weidemüller, Y.B. Ovchinnikov, in *Advances In Atomic, Molecular, and Optical Physics*, *Advances in atomic, molecular, and optical physics*, vol. 42 (Elsevier, 2000), pp. 95–170. https://doi.org/10.1016/s1049-250x(08)60186-x
6. F. Le Kien, P. Schneeweiss, A. Rauschenbeutel, Eur. Phys. J. D **67**(5), 92 (2013). https://doi.org/10.1140/epjd/e2013-30729-x
7. D. Bloch, B. Hofer, S.R. Cohen, A. Browaeys, I. Ferrier-Barbut, Phys. Rev. Lett. **131**(20), 203401 (2023). https://doi.org/10.1103/PhysRevLett.131.203401. https://link.aps.org/doi/10.1103/PhysRevLett.131.203401

8. D.S. Grün, S.J.M. White, A. Ortu, A. Di Carli, H. Edri, M. Lepers, M.J. Mark, F. Ferlaino, Phys. Rev. Lett. **133**(22), 223402 (2024). https://doi.org/10.1103/PhysRevLett.133.223402
9. L.R.B. Picard, A.J. Park, G.E. Patenotte, S. Gebretsadkan, D. Wellnitz, A.M. Rey, K.K. Ni, Nature 1–6 (2024). https://doi.org/10.1038/s41586-024-08177-3. https://www.nature.com/articles/s41586-024-08177-3
10. J.W. Lis, A. Senoo, W.F. McGrew, F. Rönchen, A. Jenkins, A.M. Kaufman, Phys. Rev. X **13**(4), 041035 (2023). https://doi.org/10.1103/PhysRevX.13.041035. https://link.aps.org/doi/10.1103/PhysRevX.13.041035
11. V. Klüsener, S. Pucher, D. Yankelev, J. Trautmann, F. Spriestersbach, D. Filin, S.G. Porsev, M.S. Safronova, I. Bloch, S. Blatt, Phys. Rev. Lett. **132**(25), 253201 (2024). https://doi.org/10.1103/physrevlett.132.253201. https://journals.aps.org/prl/abstract/10.1103/PhysRevLett.132.253201
12. S. Pucher, V. Klüsener, F. Spriestersbach, J. Geiger, A. Schindewolf, I. Bloch, S. Blatt, Phys. Rev. Lett. **132**(15), 150605 (2024). https://doi.org/10.1103/PhysRevLett.132.150605. https://journals.aps.org/prl/pdf/10.1103/PhysRevLett.132.150605
13. G. Unnikrishnan, P. Ilzhöfer, A. Scholz, C. Hölzl, A. Götzelmann, R.K. Gupta, J. Zhao, J. Krauter, S. Weber, N. Makki, H.P. Büchler, T. Pfau, F. Meinert, Phys. Rev. Lett. **132**(15), 150606 (2024). https://doi.org/10.1103/PhysRevLett.132.150606
14. T.O. Höhn, R.A. Villela, E. Zu, L. Bezzo, R.M. Kroeze, M. Aidelsburger (2024). arXiv:2412.14163
15. D.J. Wineland, W.M. Itano, Phys. Rev. A Gen. Phys. **20**(4), 1521 (1979). https://doi.org/10.1103/physreva.20.1521. https://link.aps.org/doi/10.1103/PhysRevA.20.1521
16. I.S. Madjarov, Entangling, controlling, and detecting individual strontium atoms in optical tweezer arrays. Ph.D. thesis, California Institute of Technology (2021). https://resolver.caltech.edu/CaltechTHESIS:01292021-001639979

Chapter 3
Yttebium Atoms and Rydberg States

This chapter reviews the fundamental properties of ytterbium atoms and Rydberg states. The first part summarizes the Yb energy levels and transitions utilized in this study. The second part discusses the physical characteristics of Rydberg states, the multi-channel quantum defect theory, the unique properties of two-electron Rydberg atoms, and the interactions between Rydberg atoms.

3.1 Ytterbium

3.1.1 Atomic Structure

Ytterbium (Yb) is a lanthanoid atom with an atomic number of 70. It has seven stable isotopes, five bosons (^{168}Yb, ^{170}Yb, ^{172}Yb, ^{174}Yb, ^{176}Yb) and two fermions (^{171}Yb, ^{173}Yb). The natural abundance of each isotope is summarized in Table 3.1. The electron configuration is written as [Xe]$4f^{14}6s^2$, with two valence electrons, similar to alkaline-earth atoms. This configuration gives Yb unique properties compared to alkali atoms, offering advantages for quantum computing in neutral atom arrays.

3.1.2 Energy Levels and Transitions

With two valence electrons forming spin singlet and triplet pairs, Yb atoms exhibit a greater variety of energy levels than alkali atoms. The energy levels are primarily described by the LS coupling scheme, where the total electronic orbital angular momentum L and the total electronic spin angular momentum S are coupled to form the total electronic angular momentum J, and each electronic level is denoted by $^{2S+1}L_J$. Notably, because Yb is a heavy atom, spin-orbit coupling cannot be

Y. Nakamura, *Generation of Highly Controllable Ytterbium Atom Array*,
Springer Theses, https://doi.org/10.1007/978-981-95-2836-3_3

Table 3.1 Isotopes of ytterbium

	Natural abundance (%)	Nuclear spin ($\hbar$)	μ_I/μ_N	Statistics
^{168}Yb	0.13	0		Boson
^{170}Yb	3.05	0		Boson
^{171}Yb	14.3	1/2	0.4919	Fermion
^{172}Yb	21.9	0		Boson
^{173}Yb	16.1	5/2	−0.6776	Fermion
^{174}Yb	31.8	0		Boson
^{176}Yb	12.7	0		Boson

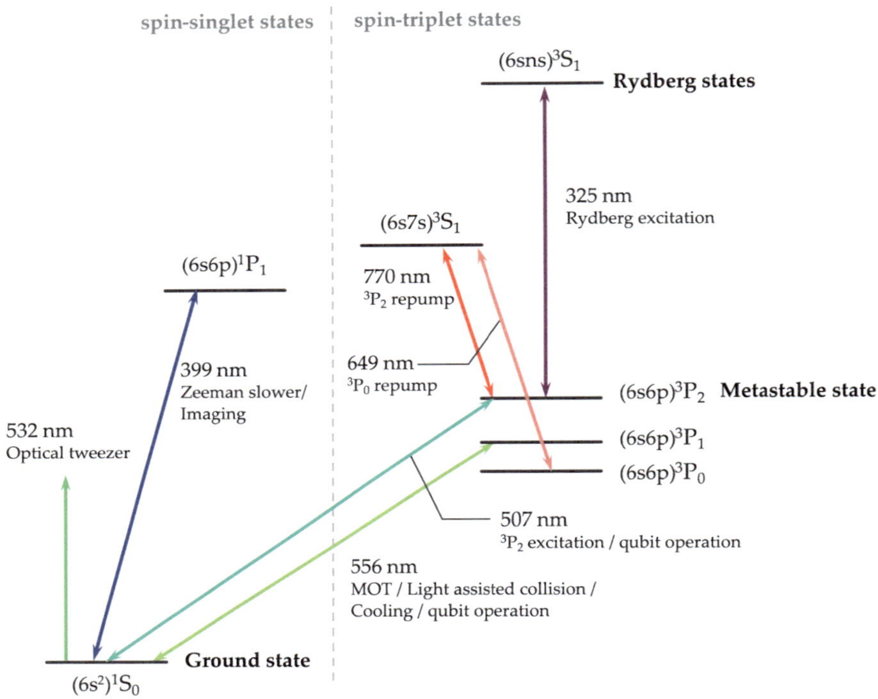

Fig. 3.1 Yb energy levels relevant to this work

ignored, leading to slight perturbations in the LS coupling scheme. As a result, the intercombination transitions $\Delta S = 1$, which is prohibited in the selection rule of the E1 transaction, is allowed. The optical transitions and energy levels used in this study are shown in Fig. 3.1. The characteristics of each transition are summarized below.

3.1.2.1 $(6s^2)^1S_0 - (6s6p)^1P_1$ Transition

This transition is an E1 transition with a wavelength $\lambda = 398.9$ nm and a natural width $\Gamma = 29$ MHz [1], making it a very strong transition. Since the $^1P_1$ state slightly decays to $(5d6s)^3D_1$ and $(5d6s)^3D_2$ [2], this transition is not completely closed. In this study, this transition is used for the Zeeman slowing and fluorescence imaging of single atoms.

3.1.2.2 $(6s^2)^1S_0 - (6s6p)^3P_1$ Transition

This transition is an intercombination transition connecting the singlet and the triplet states ($\Delta S = 1$). The transition wavelength is $\lambda = 555.8$ nm, and the natural width is $\Gamma = 182.4$ kHz [3]. Although this transition is forbidden in the E1 transtion selection rule, the heavy Yb atom has a pure LS coupling level $|^1P_1^0\rangle$, $|^3P_1^0\rangle$ due to the spin-orbit interaction as

$$\begin{aligned} |^1P_1\rangle &= \alpha\,|^1P_1^0\rangle + \beta\,|^3P_1^0\rangle, \\ |^3P_1\rangle &= \alpha\,|^3P_1^0\rangle - \beta\,|^1P_1^0\rangle, \end{aligned} \tag{3.1}$$

and the transition becomes allowed. The values of α and β are obtained from the lifetime τ and energy E of each state as

$$\frac{|\alpha|^2}{|\beta|^2} = \frac{\tau(^3P_1)E(^3P_1)^3}{\tau(^1P_1)E(^1P_1)^3}. \tag{3.2}$$

From this equation, we obtain $|\alpha|^2 = 0.9834$ and $|\beta|^2 = 0.0166$. This transition is perfectly closed transition and has a much lower Doppler cooling limit temperature of $T_D = \hbar\Gamma/(2k_B) = 4.4\,\mu$K than alkali atoms due to its narrow natural linewidth. This linewidth is narrow enough for atom cooling and simultaneously broad enough for making magneto-optical trap (MOT) without the need for other transitions [4]. Furthermore, this transition can be employed for fluorescence imaging of single atoms with enough scattering rate and low atom temperature [5, 6]. In this study we use this transition for the magneto-optical trap (MOT), Doppler cooling, light-assisted collision (LAC), Raman sideband cooling (RSC), optical pumping (OP), and single qubit gates.

Basic properties and isotope shifts of $^1S_0 \leftrightarrow {}^1P_1$ and $^1S_0 \leftrightarrow {}^3P_1$ transitions are summarized in Tables 3.2 and 3.3, respectively.

3.1.2.3 $(6s^2)^1S_0 - (6s6p)^3P_2$ Transition

This transition is M2 transition with a wavelength $\lambda = 507.35$ nm. As we discussed in Sect. 2.1.3, this transition is very weak and thus ultranarrow. In the case of fermionic isotope, the $^3P_2$ state is perturbed by hyperfine mixing as [7]

Table 3.2 Basic properties of $^1S_0 \leftrightarrow {}^1P_1$ and $^1S_0 \leftrightarrow {}^3P_1$ transitions

		$^1S_0 \leftrightarrow {}^1P_1$	$^1S_0 \leftrightarrow {}^3P_1$	Unit
Wavelength	λ	398.9	555.8	nm
Lifetime	τ	5.5	873	ns
Natural linewidth	$\Gamma/2\pi = 1/(2\pi\tau)$	29	0.182	MHz
Saturation intensity	I_{sat}	59	0.139	mW/cm^2
Recoil energy	E_R/k_B	344	178	nK
Doppler cooling limit	T_D	690	4.37	μK

Table 3.3 Isotope shifts of $^1S_0 \leftrightarrow {}^1P_1$ [1] and $^1S_0 \leftrightarrow {}^3P_1$ [3] transitions

$^1S_0 \leftrightarrow {}^1P_1$	Isotope shift (MHz)	$^1S_0 \leftrightarrow {}^3P_1$	Isotope shift (MHz)
^{168}Yb	1887.4	^{173}Yb, $(F = 3/2)$	3807.3
^{170}Yb	1192.4	^{171}Yb, $(F = 3/2)$	3804.6
^{171}Yb, $(F = 1/2)$	1153.7	^{168}Yb	3655.1
^{171}Yb, $(F = 3/2)$	832.4	^{173}Yb, $(F = 5/2)$	2311.4
^{173}Yb, $(F = 7/2)$	588.0	^{170}Yb	2286.4
^{172}Yb	533.3	^{172}Yb	1000.0
^{173}Yb, $(F = 3/2)$	516.0	^{174}Yb	0
^{174}Yb	0	^{176}Yb	−954.8
^{173}Yb, $(F = 5/2)$	−253.4	^{171}Yb, $(F = 1/2)$	−2132.1
^{176}Yb	−509.3	^{173}Yb, $(F = 7/2)$	−2386.7

$$\left|{}^3P_2, F, m_F\right\rangle = \sum_\alpha \frac{\left\langle {}^3P_2^0; F, m_F \right| \mathcal{H}_{HF} \left|\alpha\right\rangle}{E(\alpha) - E({}^3P_2^0)} \left|\alpha\right\rangle, \tag{3.3}$$

where α is the intermediate states, and the hyperfine interaction Hamiltonian $\mathcal{H}_{HF}$ is given as [8]

$$\mathcal{H}_{HF} = A\frac{\boldsymbol{I}\cdot\boldsymbol{J}}{\hbar^2} + B\frac{3(\boldsymbol{I}\cdot\boldsymbol{J})^2/\hbar^2 + 3(\boldsymbol{I}\cdot\boldsymbol{J})/(2\hbar) - I(I+1)J(J+1)}{2I(2I-1)J(2J-1)}, \tag{3.4}$$

with the hyperfine constants A and B. As a result, the electronic dipole coupling with the ground state occurs as

$$\sum_\alpha \frac{\left\langle {}^3P_2^0; F, m_F \right| \mathcal{H}_{HF} \left|\alpha\right\rangle}{E(\alpha) - E({}^3P_2^0)} \langle\alpha|\boldsymbol{d}\cdot\boldsymbol{E}|{}^1S_0; F', m_{F'}\rangle. \tag{3.5}$$

Table 3.4 Basic properties of $^1S_0 \leftrightarrow {}^3P_2$ transition[7, 9]

		Boson	^{171}Yb	^{173}Yb	Unit
Wavelength	λ		507.35		nm
Lifetime	τ	15	6.3	7.2	s
Natural linewidth	$\Gamma/2\pi = 1/(2\pi\tau)$	10.6	25	22	mHz
Saturation intensity	I_{sat}	1.1×10^{-8}	2.5×10^{-8}	2.2×10^{-8}	mW/cm^2
Recoil energy	E_R/k_B		214		nK

Table 3.5 Basic properties of $^3S_1 \leftrightarrow {}^3P_J$ transitions. The Radiative decay rate is given by $\Gamma_J = \omega_J^3 |\langle {}^3P_J || \boldsymbol{d} || {}^3S_1 \rangle|^2 / (9\pi\epsilon_0 \hbar c^3)$ with the transition angular frequency ω_J and the reduced matrix element $\langle {}^3P_J || \boldsymbol{d} || {}^3S_1 \rangle$ [9]

		$^3S_1 \rightarrow {}^3P_0$	$^3S_1 \rightarrow {}^3P_1$	$^3S_1 \rightarrow {}^3P_2$	Unit
Wavelength	λ	649.1	680.1	770.2	nm
Radiative decay rate	$\Gamma_J/(2\pi)$	1.54	4.26	6.00	MHz
Branching ratio		0.13	0.36	0.51	

This has non-zero value for $F = 3/2$ of ^{171}Yb and $F = 3/2,\ 5/2$ of ^{173}Yb, and the transition becomes hyperfine mixing induced E1 (HFM-E1) transition, with which we can obtain one order magnitude stronger Rabi frequency than the M2 transition. Table 3.4 summarizes the basic properties of the $^1S_0 \leftrightarrow {}^3P_2$ transition.

3.1.2.4 $(6s6p)^3P_2 - (6s7s)^3S_1$ Transition

This transition is E1 transition with a wavelength $\lambda = 770.2$ nm and used for $^3P_2$ state repumping. The $(6s7s)^3S_1$ state decays to not only $^3P_2$ but also $^3P_0$ and $^3P_1$ states. Therefore, we need additional repumping of the $^3P_0$ state to perfectly repump the $^3P_2$ state.

3.1.2.5 $(6s6p)^3P_0 - (6s7s)^3S_1$ Transition

This transition is E1 transition with a wavelength $\lambda = 649.1$ nm and used for $^3P_0$ state repumping. The branching ratio of the $(6s7s)^3S_1$ state is summarized in Table 3.5.

3.1.2.6 $(6s6p)^3P_2$ – (6snl) Rydberg State Transition

This transition is E1 transition with a wavelength $\lambda \simeq 325.5$ nm. The properties of the Rydberg state highly depends on the principal quantum number as we will discuss later.

3.2 Rydberg States

3.2.1 General Properties of Rydberg Atoms

Rydberg states are highly energetic states of atoms, where one of the valence electrons is excited to an orbital with a large principal quantum number n. Since the excited electron feels approximately $+e$ charge from the core, the properties of Rybderg states are in good analogy to the hydrogen atom. As the energy levels of the hydrogen atom is given by

$$E_n = -hc\frac{R_\infty}{n^2}, \tag{3.6}$$

with the Rydberg constant $R_\infty = m_e e^4/(8\epsilon_0^2 h^3 c)$ and electron mass m_e, the energy levels of the Rydberg atom is given by [10]

$$E_{n,l,J} = -hc\frac{R^*}{\nu^2}, \tag{3.7}$$

with the mass scaled Rydberg constant $R^* = R_\infty/(1 + m_e/m_c)$, core mass m_c, and the effective principal quantum number $\nu = n - \delta_{n,l,J}$. The quantum defect $\delta_{n,l,J}$ is introduced to take into account the deviation from the hydrogen atom due to the core potential, and depends on the principal quantum number n, the orbital angular momentum l, and the total angular momentum J of the Rydberg state, which is determined by the spectroscopy of the Rydberg states.

The large orbit of the Rydberg electron makes the properties of Rydberg atoms quite different from those of low-lying states. From the scaling law of the orbital radius $\langle r \rangle \sim n^2$, we can obtain the scaling law of various quantities as shown in Table 3.6.

3.2.2 Quantum Defect Theory

The quantum defect theory (QDT) is used to describe the deviation of the Rydberg wavefunction from the hydrogen atom due to the core potential. While the Rydberg electron feels approximately $+e$ charge from the core when it is far from the core, the interaction between the Rydberg electron and the core is not negligible when the

Table 3.6 Scaling laws of the properties of Rydberg states

Property		n scaling
Binding energy	E_n	n^{-2}
Energy interval	ΔE_n	n^{-3}
Orbital radius	$\langle r \rangle$	n^2
Dipole moment	$\langle g\|-er\|nl\rangle$	$n^{-3/2}$
	$\langle nl\|-er\|nl'\rangle$	n^2
Radiative lifetime	τ	n^3
Dipole interaction	C_3	n^4
Polarizability	α	n^7
van der Waals interaction	C_6	n^{11}

distance between them gets shorter. We can set this characteristic distance as r_0, and the space is divided into two regions: the outer region $r > r_0$ where the system can be treated as hydrogen-like, and the inner region $r < r_0$ where the core potential is taken into account. The QDT extract the quantities associated with the Rydberg states from the boundary condition at $r = r_0$ and $r \to \infty$. In the following, we review the (single channel) QDT for one-electron atoms and multichannel QDT (MQDT) for two-electron atoms, based on Refs. [10–17]. Note that we use atomic units in the following discussion.

3.2.2.1 Single Channel

We start the discussion with the schrödinger equation for hydrogen atom as

$$\left(-\frac{1}{2}\nabla^2 - \frac{1}{r}\right)\Psi(r,\theta,\phi) = E\Psi(r,\theta,\phi), \tag{3.8}$$

where E is the energy, r is the distance between the electron and the proton, and θ, ϕ are the angles of the spherical coordinate. The solution of the schrödinger equation is given by the spherical harmonics $Y_l^m(\theta,\phi)$ and the radial wavefunction $R(r)$ as

$$\Psi(r,\theta,\phi) = R(r)Y_l^m(\theta,\phi), \tag{3.9}$$

with the quantum number l and m. The radial wavefunction $R(r)$ satisfies the radial schrödinger equation as

$$\frac{\mathrm{d}^2R(r)}{\mathrm{d}r^2} + \frac{2}{r}\frac{\mathrm{d}^2R(r)}{\mathrm{d}r^2} - \frac{l(l+1)}{r^2}R(r) + \frac{2R(r)}{r} + 2ER(r) = 0. \tag{3.10}$$

This equation has two physically interesting solutions as

$$R(r) = \frac{f(l, E, r)}{r}, \quad \frac{g(l, E, r)}{r}, \tag{3.11}$$

where f and g are the regular and irregular Coulomb functions, respectively, which have the asymptotic form at $r \to 0$ as

$$f(l, E, r) \propto r^{l+1}, \quad g(l, E, r) \propto r^{-l}. \tag{3.12}$$

As $r \to \infty$ and $E > 0$, f and g behave as sine and cosine functions, and therefore they have phase difference of $\pi/2$. For $E < 0$, f and g can be written as

$$\begin{aligned} f &= u(l, E, r)\sin(\pi\nu) - v(l, E, r)e^{i\pi\nu}, \\ g &= -u(l, E, r)\cos(\pi\nu) + v(l, E, r)e^{i\pi\nu+1/2}, \end{aligned} \tag{3.13}$$

where $\nu = 1/\sqrt{-2E}$ and u and v are exponentially increasing and decreasing functions of r. Considering that only f is allowed as $r \to 0$ and the boundary condition at $r \to \infty$, $\sin(\pi\nu)$ in Eq. (3.13) should be zero, meaning that ν should be an integer n. Therefore, we obtain the energy levels of the hydrogen atom as $E_n = -1/(2n^2)$.

In the case of Rydberg state of alkali atoms, the core potential is not negligible. When the Rydberg electron penetrates the charge cloud of the core electrons, it feels deeper potential than $-1/r$ due to decrease of the nuclear charge shielding. As a result, the wavefunction of the Rydberg electron is pulled towards the core. Then, the radial wavefunction of the Rydberg electron $\phi(r)$ at $r > r_0$ can be written as the linear combination of f and g

$$\phi(r) \propto [f(l, E, r)\cos(\pi\delta_l) - g(l, E, r)\sin(\pi\delta_l)]/r, \tag{3.14}$$

with the radial phase shift $\pi\delta_l$. Considering the boundary condition $\phi(r) \to 0$ at $r \to \infty$, we obtain the constraint as

$$\cos(\pi\delta_l)\sin(\pi\nu) + \sin(\pi\delta_l)\cos(\pi\nu) = 0, \tag{3.15}$$

or equivalently $\sin(\pi\nu + \pi\delta_l) = 0$. This condition means that $\nu = n - \delta_l$ with an integer n, and the energy levels of the Rydberg states are given by $E_{n,l} = -1/(2\nu^2) = -1/(2(n - \delta_l)^2)$. Here, δ_l is the qunatum defect. Once the quantum defects are obtained from the spectroscopy of the Rydberg states, we can calculate the wavefunction of the Rydberg electron and various quantities associated with the Rydberg states.

3.2.2.2 Multi-channel

In contrast to the one-electron atoms, the two-electron atoms have much more complicated structure in Rydberg states due to multiple configurations of the core, which we refer as "channels".

For $r > r_0$, the radial wavefunction $\phi_i(r)$ of long-range (or collision) channel i is written as

$$\phi_i(r) = [f(l_i, E_i, r)\cos(\pi\delta_i) - g(l_i, E_i, r)\sin(\pi\delta_i)]/r, \tag{3.16}$$

where $E_i := -1/(2\nu_i^2) = E - I_i$ is the asymptotic kinetic energy of the Rydberg electron with the ionization limit I_i for ith channel, l_i is the orbital angular momentum of the Rydberg electron, and $\pi\delta_i$ is the phase shift. This equation is the counterpart of Eq. (3.14) for the one-electron atom. From the boundary condition at $r \to \infty$ as Eq. (3.15), we can replace $\delta_i = n - \nu_i$ and rewrite Eq. (3.16) as

$$\phi_i(r) = [f(l_i, E_i, r)\cos(\pi\nu_i) + g(l_i, E_i, r)\sin(\pi\nu_i)]/r. \tag{3.17}$$

The atomic wavefunction Ψ can then be written as

$$\Psi = \mathcal{A}\sum_{i=1}^{N} A_i\phi_i\chi_i, \tag{3.18}$$

where N is number of channels, A_i is the channel coefficients, χ_i is angular and spin components, and $\mathcal{A}$ is anti-symmetrized operator.

On the other hand, for $r < r_0$, the radial wavefunction $\{\phi_i\}$ is no longer good basis for the system because the singly charged core picture is invalid in this region. Instead, the wavefunction is characterized by the doubly charged core and the coupling of two valence electrons. We introduce new basis $\{\psi_\alpha\}$, where α is the short-range (or close-coupling) channel index, and the wavefunction Ψ is written as

$$\Psi = \sum_{\alpha=1}^{N} B_\alpha\psi_\alpha, \tag{3.19}$$

with the channel coefficients B_α. The basis $\{\psi_\alpha\}$ is converted to the basis $\{\psi_i\}$ for the long-range channels by a unitary transformation U. Therefore, ψ_α can be written as

$$\psi_\alpha = \frac{1}{r}\left[\mathcal{A}\sum_{i=1}^{N} U_{i\alpha} f(l, E_i, r)\chi_i\cos(\pi\mu_\alpha) - \mathcal{A}\sum_{i=1}^{N} U_{i\alpha} g(l, E_i, r)\chi_i\sin(\pi\mu_\alpha)\right], \tag{3.20}$$

where $U_{i\alpha}$ is the matrix element of U, and μ_α is eigenchannel quantum defect. From Eqs. (3.17) to (3.20), we can obtain the relation between the channel coefficients A_i and B_α as

$$\begin{aligned} A_i \cos(\pi \nu_i) &= \sum_{\alpha=1}^{N} U_{i\alpha} B_\alpha \cos(\pi \mu_\alpha), \\ A_i \sin(\pi \nu_i) &= -\sum_{\alpha=1}^{N} U_{i\alpha} B_\alpha \sin(\pi \mu_\alpha). \end{aligned} \tag{3.21}$$

Because U is unitary, we can rewrite the above equations by multiplying $U_{\alpha i}$ and taking summation over i as

$$\begin{cases} \sum_i U_{\alpha i} A_i \cos(\pi \nu_i) = B_\alpha \cos(\pi \mu_\alpha) \\ \sum_i U_{\alpha i} A_i \sin(\pi \nu_i) = -B_\alpha \sin(\pi \mu_\alpha) \end{cases} \iff \begin{cases} \sum_i U_{\alpha i} \cos(\pi(\nu_i + \mu_\alpha)) A_i = B_\alpha \\ \sum_i U_{\alpha i} \sin(\pi(\nu_i + \mu_\alpha)) A_i = 0 \end{cases} \tag{3.22}$$

for $\alpha = 1, 2, \ldots N$. These equations have non-trivial solutions only when

$$\det\left[U_{\alpha i} \sin(\pi(\nu_i + \mu_\alpha))\right] = 0 \tag{3.23}$$

is satisfied. This equation gives a constraint on the effective principal quantum number ν_i for given values of μ_α and $U_{\alpha i}$. As mentioned above, the boundary condition that $\Psi \to 0$ as $r \to 0$ leads to the additional constraint on ν_i as

$$E = I_i + E_i = I_i - \frac{1}{2\nu_i^2}. \tag{3.24}$$

By optimizing the parameters of $U_{\alpha i}$, μ_α, I_i, and ν_i in Eqs. (3.23) and (3.24) to match the spectroscopy data of the Rydberg states, we can obtain an emperical MQDT model for the Rydberg series.

When the spin-orbit interaction is considerable, a frame transformation by introducing a purly LS-coupled channel $\bar{\alpha}$ is convenient to express the collision channel i in terms of the jj-couple channel. In this case, the matrix U is decomposed as

$$U_{i\alpha} = U_{i\bar{\alpha}} V_{\bar{\alpha}\alpha}. \tag{3.25}$$

The matrix element $U_{i\bar{\alpha}}$ corresponds to the LS-jj frame transformation and is given as

$$\begin{aligned} U_{i\bar{\alpha}} &= \langle ((S_c L_c) J_c (sl) j) J | \, ((S_c s) S (L_c l) L) J \rangle \\ &= \sqrt{(2S+1)(2L+1)(2J_c+1)(2j+1)} \begin{Bmatrix} S_c & L_c & J_c \\ s & l & j \\ S & L & J \end{Bmatrix}, \end{aligned} \tag{3.26}$$

where S_c, L_c, J_c are the total spin, orbital angular momentum, and total angular momentum of the core, s, l, j are the spin, orbital angular momentum, and total angular momentum of the Rydberg electron, and S, L, J are the total spin, orbital angular momentum, and total angular momentum of all electrons. For channel i (left-hand-side of the braket), $S_c + L_c = J_c$, $s + l = j$, and $J_c + j = J$ while for channel $\bar{\alpha}$ (right-hand-side of the braket), $S_c + s = S$, $L_c + l = L$, and $S + L = J$. The curly bracket $\{\cdot\}$ in Eq. (3.26) is the Wigner 9-j symbol. The matrix element $V_{\bar{\alpha}\alpha}$ is expressed with multiple rotation matrices $R(\theta)$ as

$$V_{\bar{\alpha}\alpha} = \prod_{k=1}^{N(N-1)/2} R(\theta_k), \tag{3.27}$$

where k runs over all combinations of the jj-coupled channels $(i, j) = (1, 2), (1, 3), \ldots, (N-1, N)$, and θ_k is the Euler angle around channels i and j.

When the atom has a nuclear spin I, the hyperfine interaction should be taken into account. In this case, Eq. (3.23) is modified as

$$\det\left[U_{\alpha_F i_F} \sin(\pi(\nu_{i_F} + \mu_{\alpha_F}))\right] = 0, \tag{3.28}$$

where F represents the total angular momentum $F = J + I$. Then the matrix U is decomposed as

$$U_{i_F\alpha_F} = U_{i_F i} U_{i\alpha_F}, \tag{3.29}$$

with the transformation matrix of

$$U_{i_F i} = \langle ((J_c I) F_c j) F |\, ((J_c j) J I) F \rangle = (-1)^{I+j+F_c+J} \sqrt{(2J+1)(2F_c+1)} \begin{Bmatrix} I & J_c & F_c \\ j & F & J \end{Bmatrix}, \tag{3.30}$$

where $\{\cdot\}$ is the Wigner 6-j symbol. Since the hyperfine interaction is smaller than the non-Colombic interaction in close-coupling region, it can be treated as a perturbation. Thus, we can start to construct a MQDT model with the parameters obtained for the isotope with zero nuclear spin and Eq. (3.29).

Up to this point, the discussion has primarily focused on the matrix U. However, it is less favorable for numerical algorithms due to the presence of the sine function in Eq. (3.23). Instead, equivalent to Eq. (3.23), we can transform Eq. (3.22) as

$$\begin{aligned} \sum_i U_{\alpha i} \sin(\pi(\nu_i + \mu_\alpha)) A_i = 0 &\Longleftrightarrow \sum_i [U_{\alpha i} \tan(\pi \nu_i) + \tan(\pi \mu_\alpha) U_{\alpha i}] \cos(\pi \nu_i) A_i = 0 \\ &\Longleftrightarrow \sum_i \left[\delta_{ji} \tan(\pi \nu_i) + \sum_\alpha (U_{j\alpha} \tan(\pi \mu_\alpha) U_{\alpha i})\right] \cos(\pi \nu_i) A_i = 0 \\ &\Longleftrightarrow \sum_i \left[K_{ji} + \delta_{ji} \tan(\pi \nu_i)\right] a_i = 0, \end{aligned} \tag{3.31}$$

where we define $K := U^\dagger \mathrm{diag}(\pi \mu_\alpha) U$ and $a_i = \cos(\pi \mu_i) A_i$. Note that δ_{ji} is the Kronecker delta. The above equation has non-trivial solutions only when

$$\det\left[K_{ji} + \delta_{ji} \tan(\pi \nu_i)\right] = 0. \tag{3.32}$$

The matrix K is real symmetric matrix and referred as the reactance (or reaction) matrix, which has the information of the coupling between the channels. From

Eqs. (3.24), (3.32) and the spectroscopy data, we can estimate K matrix and ν_i, and obtain an emperical MQDT model. The frame transformation can be applied to the reaction matrix so as done to the matrix U.

3.2.3 *Two-Electron Rydberg Atoms*

In contrast to the Rydberg states of alkali atoms, the Rydberg state of two-electron atoms have an optically active inner valence electron. This offers unique properties of the two-electron Rydberg atoms: isolated core excitation (ICE) and trappable Rydberg states, which can be advantages for quantum simulation and computing. In the following, we take ytterbium atom as an example of two-electron Rydberg atoms.

3.2.3.1 Isolated Core Excitation

ICE is the transition from a $(6snl)$ Rydberg state to $(6p_{1/2}n'l')$ state [18]. This transition can be used to induce light shifts of the Rydberg states without perturbating the ground state. Combining with local addressing of the ICE laser in an optical tweezer array, we can control the Rybderg excitation locally even with global irradiation of the Rydberg excitation laser [19, 20]. The important point is that the ICE does not affect the ground state, and thus the scattering error of the qubits encoded in the ground state can be suppressed.

Another application of the ICE is auto-ionization of the Rybderg states [21]. Since the $(6p_{1/2}n'l')$ state has larger energy than the ionization limit of a single electron, the state can be connected to the continuum states $|\epsilon\rangle$. As a result, the Rydberg electron is transitioned into a continuum state by absorption of a photon, that is ionization. This process is very quick (typically, GHz order) and used for fast and high-fidelity detection of the Rydberg states.

3.2.3.2 Trappable Rydberg States

The Rydberg electron feels a potential from the oscillating electric fields of the trapping laser, which is called the ponderomotive potential and given as [22]

$$U_{\text{pond}} = \frac{1}{2\epsilon_0 c}\frac{e^2}{m\omega^2}\int \mathrm{d}^3r\, |\psi_{nl}(\boldsymbol{r})|^2\, I\,(\boldsymbol{r}+\boldsymbol{R})\,, \tag{3.33}$$

where ω is the angular frequency of the trapping light, $\boldsymbol{r}$ is the Rydberg electron position with respect to the nucleus $\boldsymbol{R}$, $I(\boldsymbol{r})$ is trapping light intensity, and $\psi_{nl}(\boldsymbol{r})$ is the Rydberg electron wavefunction. This potential is always positive, and therefore the Rydberg electron feels a repulsive force. This is the reason why the Rydberg atoms of alkali are not trappable with commonly used red-detuned dipole traps. For this, the traps are usually turned off during the Rydberg excitation, limiting the interaction time between the Rybderg atoms and leading a risk to heat and lose the atoms [23].

On the other hand, in the case of two-electron Rydberg atoms, the ionic core has a finite polarizability α_{ion}. In such a situation, the light shift is written as

$$U_{\text{total}} = -\frac{1}{2\epsilon_0 c}\alpha_{\text{ion}} I(\boldsymbol{R}) + \frac{1}{2\epsilon_0 c}\frac{e^2}{m\omega^2}\int d^3r \, |\psi_{nl}(\boldsymbol{r})|^{2} I(\boldsymbol{r}+\boldsymbol{R}) \, . \tag{3.34}$$

For large n, the overlap of the Rydberg electron wavefunction and the trapping laser is small, and the second term is smaller than the first term, resulting in the attractive potential [24]. Therefore one can fully utilize the Rydberg lifetime and suppress the heating and loss of the Rydberg atoms.

3.2.4 Rydberg Interaction and Entanglement Generation

Due to the large dipole moment of the Rydberg states, the atoms in Rydberg states can interact with each other even at a distance of several micrometers. This interaction is used to generate entanglement between the atoms in quantum computation [25, 26]. The interaction is described by the Hamiltonian given as

$$\mathcal{V}_{\text{dd}} = \frac{1}{4\pi\epsilon_0}\frac{\boldsymbol{d}_1 \cdot \boldsymbol{d}_2 - 3(\boldsymbol{d}_1 \cdot \boldsymbol{R}/R)(\boldsymbol{d}_2 \cdot \boldsymbol{R}/R)}{R^3}, \tag{3.35}$$

where $\boldsymbol{d}_1$ and $\boldsymbol{d}_2$ are the dipole moments of the atoms, and $\boldsymbol{R}$ is the distance vector between the atoms. Let us consider the atoms in the same Rydberg state $|nj\rangle$, where n is the principal quantum number and j is the other quantum number of the Rydberg state. The energy shift of the pair state $|nj; nj\rangle$ can be calculated by the perturbation theory as

$$E_{\text{dd}}^{(2)} = \sum_{n_1 j_1 n_2 j_2} \frac{|\langle nj; nj|V_{\text{dd}}|n_1 j_1; n_2 j_2\rangle|^2}{\Delta E}, \tag{3.36}$$

with $\Delta E = 2E_{nj} - (E_{n_1 j_1} + E_{n_2 j_2})$. Here n_1, j_1, n_2, j_2 run over all possible Ryderg states except for $|nj\rangle$, and E_{nl} represents the unperturbed energy of the Rydberg state. Note that the first order perturbation term $\langle nj; nj|V_{\text{dd}}|nj; nj\rangle$ vanishes because the interaction Hamiltonian of Eq. (3.35) has odd parity. From Table 3.6, one can find the scaling law of the Rydberg interaction as

$$E_{\text{dd}}^{(2)} = \frac{C_6}{R^6} \sim \frac{1}{R^6}\frac{\left(n^2 \cdot n^2\right)^2}{n^{-3}} = \frac{n^{11}}{R^6}. \tag{3.37}$$

This interaction is proportional to R^{-6} and is called van der Waals interaction. In optical tweezer arrays, the matrix element $|\langle nj; nj\,|V_{\rm dd}|\,n_1 j_1; n_2 j_2\rangle|^2$ is typically smaller than the energy difference ΔE due to relatively large site intervals of several micrometers, and thus the Rydberg interaction becomes van der Waals type.

However, ΔE sometimes becomes close to zero coincidentally for a certain Rydberg state, which is called Förster resonance. In this situation, the first order perturbation does not vanishes due to the energy degeneracy, and the Rydberg interaction becomes

$$E^{(1)}_{\rm dd} = \frac{C^3}{R^3} \sim \frac{n^4}{R^3}. \tag{3.38}$$

This interaction is called resonant dipole-dipole interaction and stronger than the van der Waals interaction. The Förster resonance is found in $|59D_{3/2}59D_{3/2}\rangle \leftrightarrow |61P_{1/2}57F_{5/2}\rangle$ of Rubidium system [27], for example. Recently, Förster resonances due to the ionic core hyperfine splitting of the two-electron atoms are found in ytterbium and strontium system [16, 17]. Note that in addition to the intrinsic Förster resonance, the electric field can induce the resonance by tuning the energy levels of the Rydberg states, which is called Stark-tuned Förster resonance [10, 28, 29].

The dipole-dipole interaction has angular dependence. The interaction Hamiltonian is decomposed into the spherical basis as [30]

$$\begin{aligned} \mathcal{V}_{\rm dd} = \frac{1}{4\pi\epsilon_0 R^3}\big[&\mathcal{A}_1(\theta)\,(d_{1+}d_{2-} + d_{1-}d_{2+} + 2d_{1z}d_{2z}) \\ &+ \mathcal{A}_2(\theta)\,(d_{1+}d_{2z} - d_{1-}d_{2z} + d_{1z}d_{2+} - d_{1z}d_{2-}) \\ &+ \mathcal{A}_3(\theta)\,(d_{1+}d_{2+} - d_{1-}d_{2-})\big], \end{aligned} \tag{3.39}$$

where $d_{k,\pm} = \mp(d_{k,x} \pm i d_{k,y})$ are the spherical components of the dipole moment, and θ is the angle between the distance vector $\boldsymbol{R}$ and the quantization axis. The coefficients $\mathcal{A}_i(\theta)$ are given as

$$\mathcal{A}_1(\theta) = \frac{1-3\cos^2\theta}{2}, \qquad \mathcal{A}_2(\theta) = \frac{3\sin\theta\cos\theta}{\sqrt{2}}, \qquad \mathcal{A}_3(\theta) = \frac{3\sin^2\theta}{2}. \tag{3.40}$$

The index i of $\mathcal{A}_i(\theta)$ corresponds to the difference of the magnetic quantum number between the pair states ($\Delta m = 0,\ \pm 1,\ \pm 2$). Note that, in the case of van der Waals interaction, the angular dependence is almost averaged out in the summation of the pair states, and the interaction is nearly isotropic unless the pair states are close to the Förster resonance.

3.2.4.1 Rydberg Blockade

When the Rydberg interaction strength $V_{\rm dd}$ is larger than the Rybderg excitation Rabi frequency Ω, the simultaneous Rydberg excitation is suppressed (Fig. 3.2).

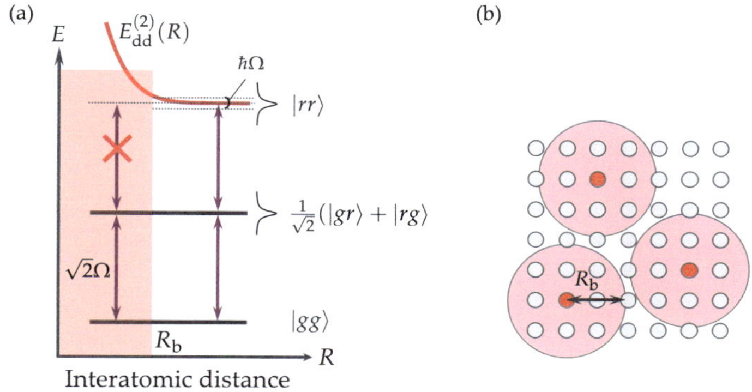

Fig. 3.2 Rydberg blockade. **a** The Rydberg excitation of a second atom is suppressed due to the energy shift caused by the Rydberg interaction. **b** Rydberg blockade in an atom array. Only a single atom can be excited to the Rydberg state within the blockade radius

This phenomenon is called Rydberg blockade [31]. Since the interaction strength depends on the interatomic distance, we can define the blockade radius $R_{\rm b}$ as the distance where the interaction strength equals to the Rabi frequency as

$$R_{\rm b} = \left(\frac{C_6}{\hbar\Omega}\right)^{1/6}. \tag{3.41}$$

More quantitatively, the Hamiltonian of two atoms under irradiation of the Rydberg excitation laser is given as

$$\begin{aligned}\mathcal{H} &= \frac{\hbar\Omega}{2}\left(|r\rangle\langle g| \otimes I + I \otimes |r\rangle\langle g| + \text{h.c.}\right) + V_{\rm dd}\,|rr\rangle\langle rr| \\ &= \hbar\frac{\sqrt{2}\Omega}{2}\left(\frac{|gr\rangle + |rg\rangle}{\sqrt{2}}\langle gg| + |rr\rangle\frac{\langle rg| + \langle gr|}{\sqrt{2}} + \text{h.c.}\right) + V_{\rm dd}\,|rr\rangle\langle rr|,\end{aligned} \tag{3.42}$$

where $|g\rangle$ and $|r\rangle$ are the ground and Rydberg states, respectively, and $I = |g\rangle\langle g| + |r\rangle\langle r|$ is the identity operator. While the first term and its hermite conjugate represent the Rabi oscillation between $|gg\rangle$ and $|\psi\rangle = (|gr\rangle + |rg\rangle)/\sqrt{2}$ with zero detuning, the second, third term, and their hermite conjugate represent the Rabi oscillation between $|\psi\rangle$ and $|rr\rangle$ with a detuning of $V_{\rm dd}$. Within the blockade radius, where the Rabi frequency is smaller than the interaction strength, the simultaneous excitation is blocked due to large detuning. As a result, the state of the two atoms oscillates between the ground state $|gg\rangle$ and the entangle state $|\psi\rangle$ with a frequency of $\sqrt{2}\Omega$. Similarly, when N atoms are in the blockade radius, the state of the system oscillates as

$$|g_1 g_2 \dots g_N\rangle \leftrightarrow \frac{1}{\sqrt{N}} \sum_{i=1}^{N} |g_1 g_2 \dots r_i \dots g_N\rangle\,, \tag{3.43}$$

with a frequency of $\sqrt{N}\Omega$.

3.2.4.2 Implementation of Controlled-Z Gate via Rydberg Interaction

The most important application of the Rydberg interaction in the optical tweezer arrays is implementation of the controlled-Z (CZ) gates between the qubits, which are essential in universal quantum computation. In the following, we review the Rydberg interaction mediated CZ gates.

The first CZ gate is proposed in 2000 [25] and experimentally demonstrated in 2010 [32–34]. The gate scheme is based on local addressing of the Rydberg excitation laser and the Rydberg blockade. Let us consider two atoms, one of which is "control" qubit and the other is "target" qubit. The atoms have qubit subspace $|0\rangle$ and $|1\rangle$, and the Rydberg state $|r\rangle$. First, a π-pulse that transfers $|1\rangle$ to $|r\rangle$ is applied to the control qubit. Then, a 2π-pulse is applied to the target qubits. If the control qubit is initially in $|1\rangle$, the Rydberg excitation is suppressed due to the Rydberg blockade, and the state of the target qubit remains unchanged. On the other hand, if the control qubit is initially in $|0\rangle$, the Rydberg excitation of the target qubit occurs, and the state of the target qubit obtains a phase of -1. Finally, a π-pulse is applied to the control qubit. This sequence of the pulses changes the state as

$$\begin{array}{llll}
|0\rangle_c|0\rangle_t \longrightarrow & |0\rangle_c|0\rangle_t \longrightarrow & |0\rangle_c|0\rangle_t \longrightarrow & |0\rangle_c|0\rangle_t \\
|0\rangle_c|1\rangle_t \longrightarrow & |0\rangle_c|1\rangle_t \longrightarrow & -|0\rangle_c|1\rangle_t \longrightarrow & -|0\rangle_c|1\rangle_t \\
|1\rangle_c|0\rangle_t \longrightarrow & i\,|r\rangle_c|0\rangle_t \longrightarrow & i\ |r\rangle_c|0\rangle_t \longrightarrow & -|1\rangle_c|0\rangle_t \\
|1\rangle_c|1\rangle_t \longrightarrow & i\,|r\rangle_c|1\rangle_t \longrightarrow & i\ |r\rangle_c|1\rangle_t \longrightarrow & -|1\rangle_c|1\rangle_t
\end{array}$$

where the subscript c and t represent the control and target qubits, respectively. One can find that this state change is equivalent to the CZ gate by relabeling the states $|0\rangle$ and $|1\rangle$, up to global phase.

The above scheme relies on local control of the Rydberg excitation laser. However, this is technically demanding in the optical tweezer arrays. To overcome this difficulty, the CZ gate with global irradiation of Rydberg excitation laser is demonstrated in 2019 [35]. This scheme uses two Rydberg pulses with a certain detuning and relative phase jump between the pulses. While the dynamics of the state $|01\rangle$ and $|10\rangle$ are given by the single atom coupling of $|1\rangle \leftrightarrow |r\rangle$ with a Rabi frequency of Ω and detuning of Δ, the state $|11\rangle$ evolvs with the collective coupling of $|11\rangle \leftrightarrow |\psi\rangle = (|1r\rangle + |r1\rangle)/\sqrt{2}$ with a Rabi frequency of $\sqrt{2}\Omega$ due to the Rybderg blockade. For a given detuning Δ, one can choose the pulse width τ such that the pulse becomes 2π-pulse for the state $|11\rangle$, and the phase jump of Rabi frequency $\Omega \rightarrow \Omega e^{i\xi}$ such that the second pulse returns the state $|01\rangle$ and $|10\rangle$ to the initial state up to phase shift. This sequence changes the states as

$$
\begin{aligned}
|00\rangle &\rightarrow |00\rangle\,, \\
|01\rangle &\rightarrow |01\rangle\, e^{i\phi}, \\
|10\rangle &\rightarrow |10\rangle\, e^{i\phi}, \\
|11\rangle &\rightarrow |11\rangle\, e^{i(2\phi-\pi)},
\end{aligned}
\tag{3.44}
$$

where ϕ is the phase obtained during the dynamics, which correspond to the geometric surface area of the Bloch sphere enclosed by the trajectory of the state. This state change is equivalent to the CZ gate up to global phase.

More recently, the CZ gate with the global irradiation of single pulse which has amplitude and phase modulation is proposed [36]. This gate scheme is called time-optimal CZ gate, and the modulation is optimized by the gradient ascent pulse engineering (GRAPE) technique. This gate is expected to be more robust against the finite Rydberg blockade strength, finite Rydberg lifetime, and experimental imperfection such as the fluctuation of the Rabi frequency and detuning. With this scheme, gate fidelity around 0.995 is achieved [16, 37–40]. The highest fidelity of the state of the art is 0.9972(3) demonstrated in ytterbium system [41].

References

1. D. Das, S. Barthwal, A. Banerjee, V. Natarajan, Phys. Rev. A **72**(3), 032506 (2005). https://link.aps.org/doi/10.1103/PhysRevA.72.032506
2. K. Honda, Y. Takahashi, T. Kuwamoto, M. Fujimoto, K. Toyoda, K. Ishikawa, T. Yabuzaki, Phys. Rev. A **59**(2), R934 (1999). https://doi.org/10.1103/physreva.59.r934. https://journals.aps.org/pra/abstract/10.1103/PhysRevA.59.R934
3. K. Pandey, A.K. Singh, P.V.K. Kumar, M.V. Suryanarayana, V. Natarajan, Phys. Rev. A **80**(2), 022518 (2009). https://link.aps.org/doi/10.1103/PhysRevA.80.022518
4. T. Kuwamoto, K. Honda, Y. Takahashi, T. Yabuzaki, Phys. Rev. A **60**(2), R745 (1999). https://doi.org/10.1103/physreva.60.r745. https://journals.aps.org/pra/abstract/10.1103/PhysRevA.60.R745
5. S. Saskin, J.T. Wilson, B. Grinkemeyer, J.D. Thompson, Phys. Rev. Lett. **122**(14), 143002 (2019). http://dx.doi.org/10.1103/PhysRevLett.122.143002
6. D. Okuno, Y. Nakamura, T. Kusano, Y. Takasu, N. Takei, H. Konishi, Y. Takahashi, J. Phys. Soc. Jpn. **91**(8), 084301 (2022). https://doi.org/10.7566/JPSJ.91.084301
7. S.G. Porsev, A. Derevianko, Phys. Rev. A **69**(4) (2004). https://doi.org/10.1103/physreva.69.042506. https://journals.aps.org/pra/abstract/10.1103/PhysRevA.69.042506
8. D.A. Steck, *Quantum and Atom Optics* (2007). https://atomoptics.uoregon.edu/~dsteck/teaching/quantum-optics/
9. S.G. Porsev, Y.G. Rakhlina, M.G. Kozlov, Phys. Rev. A **60**(4), 2781 (1999). https://doi.org/10.1103/physreva.60.2781. https://journals.aps.org/pra/abstract/10.1103/PhysRevA.60.2781
10. T.F. Gallagher, *Rydberg Atoms*. Cambridge Monographs on Atomic, Molecular and Chemical Physics (Cambridge University Press, 1994). https://www.cambridge.org/core/books/rydberg-atoms/B610BDE54694936F496F59F326C1A81B
11. M.J. Seaton, Rep. Prog. Phys. **46**(2), 167 (1983). https://doi.org/10.1088/0034-4885/46/2/002. https://iopscience.iop.org/article/10.1088/0034-4885/46/2/002/meta
12. M. Aymar, C.H. Greene, E. Luc-Koenig, Rev. Mod. Phys. **68**(4), 1015 (1996). https://doi.org/10.1103/revmodphys.68.1015. http://link.aps.org/pdf/10.1103/RevModPhys.68.1015

13. O. Robaux, M. Aymar, Comput. Phys. Commun. **25**(3), 223 (1982). https://doi.org/10.1016/0010-4655(82)90019-4. http://dx.doi.org/10.1016/0010-4655(82)90019-4
14. S. Ross, in *AIP Conference Proceedings*, vol. 225 (AIP, 1991), pp. 73–110. https://doi.org/10.1063/1.40550. https://pubs.aip.org/aip/acp/article-pdf/225/1/73/11605627/73_1_online.pdf
15. F. Robicheaux, D.W. Booth, M. Saffman, Phys. Rev. A **97**(2), 022508 (2018). https://doi.org/10.1103/PhysRevA.97.022508. https://link.aps.org/doi/10.1103/PhysRevA.97.022508
16. M. Peper, Y. Li, D.Y. Knapp, M. Bileska, S. Ma, G. Liu, P. Peng, B. Zhang, S.P. Horvath, A.P. Burgers, J.D. Thompson, Phys. Rev. X. **15**(1), 011009 (2025). https://doi.org/10.1103/physrevx.15.011009. http://link.aps.org/pdf/10.1103/PhysRevX.15.011009
17. F. Hummel, S. Weber, J. Mögerle, H. Menke, J. King, B. Bloom, S. Hofferberth, M. Li, Phys. Rev. A **110**(4), 042821 (2024). https://doi.org/10.1103/physreva.110.042821. http://link.aps.org/pdf/10.1103/PhysRevA.110.042821
18. H. Lehec, X. Hua, P. Pillet, P. Cheinet, Phys. Rev. A **103**(2), 022806 (2021). https://doi.org/10.1103/PhysRevA.103.022806. https://link.aps.org/doi/10.1103/PhysRevA.103.022806
19. K.L. Pham, T.F. Gallagher, P. Pillet, S. Lepoutre, P. Cheinet, PRX Quantum **3**(2), 020327 (2022). https://doi.org/10.1103/PRXQuantum.3.020327. https://link.aps.org/doi/10.1103/PRXQuantum.3.020327
20. A.P. Burgers, S. Ma, S. Saskin, J. Wilson, M.A. Alarcón, C.H. Greene, J.D. Thompson, PRX Quantum **3**(2), 020326 (2022). https://doi.org/10.1103/PRXQuantum.3.020326. https://link.aps.org/doi/10.1103/PRXQuantum.3.020326
21. I.S. Madjarov, J.P. Covey, A.L. Shaw, J. Choi, A. Kale, A. Cooper, H. Pichler, V. Schkolnik, J.R. Williams, M. Endres, Nat. Phys. **16**(8), 857 (2020). https://doi.org/10.1038/s41567-020-0903-z. https://www.nature.com/articles/s41567-020-0903-z
22. I.S. Madjarov, J.P. Covey, A.L. Shaw, J. Choi, A. Kale, A. Cooper, H. Pichler, V. Schkolnik, J.R. Williams, M. Endres, Nat. Phys. **16**(8), 857 (2020). https://doi.org/10.1038/s41567-020-0903-z. https://www.nature.com/articles/s41567-020-0903-z
23. I.S. Madjarov, J.P. Covey, A.L. Shaw, J. Choi, A. Kale, A. Cooper, H. Pichler, V. Schkolnik, J.R. Williams, M. Endres, Nat. Phys. **16**(8), 857 (2020). https://doi.org/10.1038/s41567-020-0903-z. https://www.nature.com/articles/s41567-020-0903-z
24. I.S. Madjarov, J.P. Covey, A.L. Shaw, J. Choi, A. Kale, A. Cooper, H. Pichler, V. Schkolnik, J.R. Williams, M. Endres, Nat. Phys. **16**(8), 857 (2020). https://doi.org/10.1038/s41567-020-0903-z. https://www.nature.com/articles/s41567-020-0903-z
25. D. Jaksch, J.I. Cirac, P. Zoller, S.L. Rolston, R. Cote, M.D. Lukin, Phys. Rev. Lett. **85**(10), 2208 (2000). https://doi.org/10.1103/PhysRevLett.85.2208. http://dx.doi.org/10.1103/PhysRevLett.85.2208
26. M.D. Lukin, M. Fleischhauer, R. Cote, L.M. Duan, D. Jaksch, J.I. Cirac, P. Zoller, Phys. Rev. Lett. **87**(3), 037901 (2001). https://doi.org/10.1103/PhysRevLett.87.037901. http://link.aps.org/pdf/10.1103/PhysRevLett.87.037901
27. S. Ravets, H. Labuhn, D. Barredo, L. Béguin, T. Lahaye, A. Browaeys, Nat. Phys. **10**(12), 914 (2014). https://doi.org/10.1038/nphys3119. https://www.nature.com/articles/nphys3119
28. I.I. Ryabtsev, D.B. Tretyakov, I.I. Beterov, V.M. Entin, Phys. Rev. Lett. **104**(7), 073003 (2010). https://doi.org/10.1103/PhysRevLett.104.073003. https://journals.aps.org/prl/abstract/10.1103/PhysRevLett.104.073003
29. J. Nipper, J.B. Balewski, A.T. Krupp, B. Butscher, R. Löw, T. Pfau, Phys. Rev. Lett. **108**(11), 113001 (2012). https://doi.org/10.1103/PhysRevLett.108.113001. https://journals.aps.org/prl/pdf/10.1103/PhysRevLett.108.113001
30. S. Ravets, H. Labuhn, D. Barredo, T. Lahaye, A. Browaeys, Phys. Rev. A **92**(2) (2015). https://doi.org/10.1103/physreva.92.020701. https://journals.aps.org/pra/pdf/10.1103/PhysRevA.92.020701
31. M. Saffman, T.G. Walker, K. Mølmer, Rev. Mod. Phys. **82**(3), 2313 (2010). https://doi.org/10.1103/RevModPhys.82.2313. https://link.aps.org/doi/10.1103/RevModPhys.82.2313
32. L. Isenhower, E. Urban, X.L. Zhang, A.T. Gill, T. Henage, T.A. Johnson, T.G. Walker, M. Saffman, Phys. Rev. Lett. **104**(1), 010503 (2010). https://doi.org/10.1103/PhysRevLett.104.010503. http://link.aps.org/pdf/10.1103/PhysRevLett.104.010503

33. X.L. Zhang, L. Isenhower, A.T. Gill, T.G. Walker, M. Saffman, Phys. Rev. A **82**(3), 030306 (2010). https://doi.org/10.1103/PhysRevA.82.030306. https://link.aps.org/doi/10.1103/PhysRevA.82.030306
34. T. Wilk, A. Gaëtan, C. Evellin, J. Wolters, Y. Miroshnychenko, P. Grangier, A. Browaeys, Phys. Rev. Lett. **104**(1), 010502 (2010). https://doi.org/10.1103/PhysRevLett.104.010502. https://journals.aps.org/prl/pdf/10.1103/PhysRevLett.104.010502
35. H. Levine, A. Keesling, G. Semeghini, A. Omran, T.T. Wang, S. Ebadi, H. Bernien, M. Greiner, V. Vuletić, H. Pichler, M.D. Lukin, Phys. Rev. Lett. **123**(17), 170503 (2019). https://doi.org/10.1103/PhysRevLett.123.170503. http://dx.doi.org/10.1103/PhysRevLett.123.170503
36. S. Jandura, G. Pupillo, Quantum **6**(712), 712 (2022). https://doi.org/10.22331/q-2022-05-13-712. https://quantum-journal.org/papers/q-2022-05-13-712/pdf/
37. S.J. Evered, D. Bluvstein, M. Kalinowski, S. Ebadi, T. Manovitz, H. Zhou, S.H. Li, A.A. Geim, T.T. Wang, N. Maskara, H. Levine, G. Semeghini, M. Greiner, V. Vuletić, M.D. Lukin, Nature (London) **622**(7982), 268 (2023). https://doi.org/10.1038/s41586-023-06481-y. urlhttp://dx.doi.org/10.1038/s41586-023-06481-y
38. P. Scholl, A.L. Shaw, R.B.S. Tsai, R. Finkelstein, J. Choi, M. Endres, Nature (London) **622**(7982), 273 (2023). https://doi.org/10.1038/s41586-023-06516-4. https://www.nature.com/articles/s41586-023-06516-4
39. A. Cao, W.J. Eckner, T. Lukin Yelin, A.W. Young, S. Jandura, L. Yan, K. Kim, G. Pupillo, J. Ye, N. Darkwah Oppong, A.M. Kaufman, Nature **634**(8033), 315 (2024). https://doi.org/10.1038/s41586-024-07913-z. https://www.nature.com/articles/s41586-024-07913-z
40. R.B.S. Tsai, X. Sun, A.L. Shaw, R. Finkelstein, M. Endres, PRX quantum **6**(1), 010331 (2025). https://doi.org/10.1103/prxquantum.6.010331. http://link.aps.org/pdf/10.1103/PRXQuantum.6.010331
41. J.A. Muniz, M. Stone, D.T. Stack, M. Jaffe, J.M. Kindem, L. Wadleigh, E. Zalys-Geller, X. Zhang, C.A. Chen, M.A. Norcia, J. Epstein, E. Halperin, F. Hummel, T. Wilkason, M. Li, K. Barnes, P. Battaglino, T.C. Bohdanowicz, G. Booth, A. Brown, M.O. Brown, W.B. Cairncross, K. Cassella, R. Coxe, D. Crow, M. Feldkamp, C. Griger, A. Heinz, A.M.W. Jones, H. Kim, J. King, K. Kotru, J. Lauigan, J. Marjanovic, E. Megidish, M. Meredith, M. McDonald, R. Morshead, S. Narayanaswami, C. Nishiguchi, T. Paule, K.A. Pawlak, K.L. Pudenz, D.R. Pérez, A. Ryou, J. Simon, A. Smull, M. Urbanek, R.J.M. van de Veerdonk, Z. Vendeiro, T.Y. Wu, X. Xie, B.J. Bloom (2024). arXiv:2411.11708

Chapter 4
Experimental Apparatus and Techniques

This chapter reviews the apparatus and techniques used in the experiments. The chapter is divided into two sections. The first section describes the experimental apparatus, including the vacuum system, the magnetic field system, the laser system, and the optical setup. The second section describes the experimental techniques, including the optical tweezer array generation and trapping and imaging of single ytterbium atoms.

4.1 Experimental Apparatus

4.1.1 Vacuum System

Our vacuum chamber consists of four parts: oven chamber, Zeeman slower pipe, science chamber, and pumping chamber (Fig. 4.1). The oven chamber is used to heat the ytterbium oven to produce ytterbium gas. The generated Yb gas is then cooled by Zeeman slowing techineque in the Zeeman slower pipe. The pre-cooled Yb atoms are trapped and further cooled by magneto-optical trap (MOT) via ${}^1S_0$–${}^3P_1$ transtion in the science chamber. The pumping chamber has an ion pump to maintain the vacuum in the science chamber. With the aid of additional ion pump and titanium sublimation pumps, the pressure at the science chamber is maintained at approximately 3×10^{11} Torr.

The science chamber is specially designed for the optical tweezer array experiment and Rydberg spectroscopy (Fig. 4.2). The chamber is based on a commercial chamber (Kimball Physics, MCF800-ExtOct-G2C8A16), which has two ICF203, eight ICF70 and sixteen ICF34 flanges. Two reentrant viewports are installed on the top and bottom of the chamber (ICF203 flanges) for hosting the objective lenses, which are used for generating optical tweezers and imaging single atoms. Eight electrodes and a micro-channel plate (Hamamatsu Photonics, F4655-11) are installed inside

Y. Nakamura, *Generation of Highly Controllable Ytterbium Atom Array*,
Springer Theses, https://doi.org/10.1007/978-981-95-2836-3_4

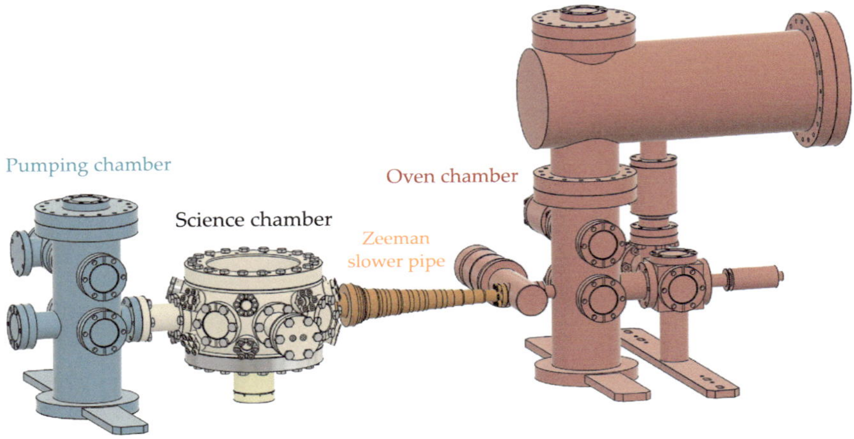

Fig. 4.1 Overview of the vacuum chamber

Fig. 4.2 Schematic of the science chamber. The chamber has two reentrant (black) for the objective lenses, eight electrodes for the electric field control (yellow), and a micro-channel plate for the Rydberg atom detection

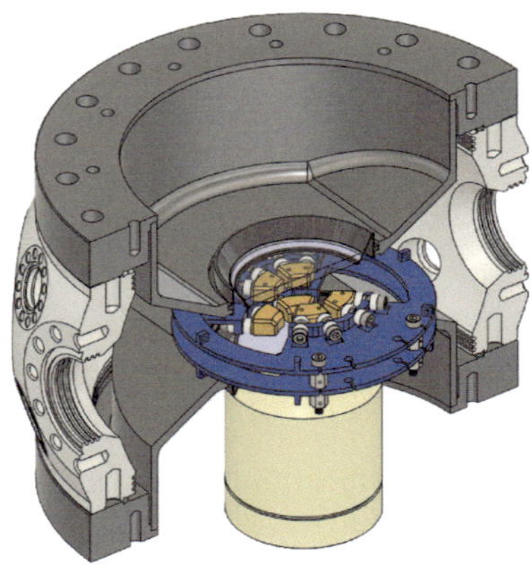

the chamber for controlling the electric field and detecting the Rydberg atoms. The windows of reentrant viewports are indium-tin-oxcide (ITO) coated to avoid charging up and to provide a good stability of the electric fields at the atom position. The details of the electrodes and the micro-channel plate are described in Ref. [1].

4.1.2 Magnetic Field System

There are four coils around the science chamber: the MOT coil and three compensation coils for x, y, and z direction. The MOT coil is used to generate the magnetic field gradient for the magneto-optical trap with anti-Helmholtz configuration, and

Table 4.1 Specifications of the coils

	Number of turns	Configuration	Magnetic field
MOT coil	42 × 14	Anti-Helmholtz	11.6 (G/cm)/A
		Helmholtz	42.7G/A
xy compensation	70	Helmholtz	0.5 G/A
z compensation	40	Helmholtz	3.4 G/A

strong bias field along z direction with Helmholtz configuration. The compensation coils are used to control the magnetic field at the atom position. Table 4.1 shows the specifications of the coils.

4.1.3 Laser System

4.1.3.1 399 nm Lasers for Probe

We have two 399 nm lasers for the probe of single atoms. The first laser is a commercial laser (Toptica, DL pro) with a power up to 70 mW. The second laser is a homemade interference-filter based external-cavity diode laser (IFDL) with a power of approximately 10 mW. The formar is used for the probe of ^{171}Yb and the latter is for ^{174}Yb. These are frequency stabilized with a wavemeter (HighFinesse, WS8-10).

4.1.3.2 399 nm Laser for Zeeman Slowing

The Zeeman slower laser is generated by a intra-cavity second harmonic generation (SHG) from 798 nm light. The fundamental light with a power of 1.5 W from a commercial laser (Toptica, TA pro) is sent to a homemade SHG cavity with a litium triborate (LBO) crystal. The cavity has a free spectral range (FSR) and finesse of 420 MHz and 380, respectively. The SHG output power is approximately 250 mW and delivered to the Zeeman slower through an end-cap fiber. The frequency of the fundamental light is stabilized a wavemeter.

4.1.3.3 507 nm Laser for Metastable State Excitation

The 507 nm laser is consisted of a commercial 1014 nm IFDL (Optoquest), a homemade tapered amplifier (TA), and a SHG with a periodically poled lithium niobate (PPLN) crystal (NTT electronics). Since the $^1S_0$–$^3P_2$ transition is ultranarrow, the 1014 nm IFDL is frequency stabilized with a ultra-low expansion (ULE) cavity with a FSR of 1.5 GHz and a finesse of 5.8×10^5, resulting in a laser linewidth of approximately 20 Hz, measured by a beat note with a stabilized optical frequency comb. The output power of the 507 nm laser is approximately 120 mW.

4.1.3.4 532 nm Laser for Optical Tweezers

The optical tweezers are generated by a 532 nm diode-pumped solid-state laser (Coherent, Verdi V-10) with a power of approximately 10 W. The polarizabilities of the $^1S_0$, $^3P_1$, and $^3P_2$ states of ^{174}Yb for this wavelength are summarized in Table 4.2. From these values, we can find that this wavelength can satisfies the near-magic conditions for the $|^1S_0\rangle \leftrightarrow |^3P_1, m_J = 0\rangle$ and $|^1S_0\rangle \leftrightarrow |^3P_2, m_J = 0\rangle$ transitions for ^{174}Yb as shown in Fig. 4.3, enabling efficient cooling and narrow linewidth excitation.

Table 4.2 Scalar and tensor polarizabilities of the $^1S_0$ [2], $^3P_1$ [2], and $^3P_2$ [3] states of ^{174}Yb at 532 nm. The polarizabilities are given in Hz/(W/cm^2) and the light shift can be calculated by Eq. 2.68

State	α^S_{nJ}	α^T_{nJ}
$^1S_0$	37.9	–
$^3P_1$	22.4(2)	–7.6(1)
$^3P_2$	46.0(2)	7.5(3)

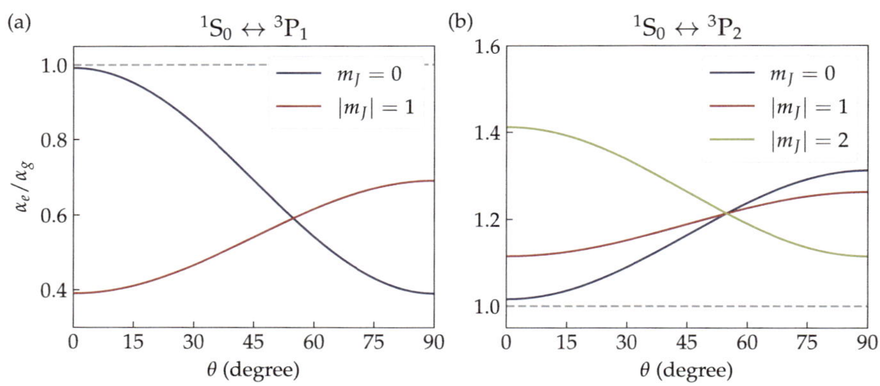

Fig. 4.3 Polarizabilities of the $^1S_0$, $^3P_1$, and $^3P_2$ states of ^{174}Yb at 532 nm. **a** The ratio of the polarizabilities of the $^1S_0$ (α_g) and the $^3P_1$ (α_e) states. The $^3P_1$ atoms feel shallower trap than that for $^1S_0$ because $\alpha_e/\alpha_g < 1$. **b** The ratio of the polarizabilities of the $^1S_0$ (α_g) and the $^3P_2$ (α_e) states. The $^3P_2$ atoms feel deeper trap than that for $^1S_0$ because $\alpha_e/\alpha_g > 1$. The near-magic conditions for the $|^1S_0\rangle \leftrightarrow |^3P_1, m_J = 0\rangle$ and $|^1S_0\rangle \leftrightarrow |^3P_2, m_J = 0\rangle$ transitions are satisfied when the tweezer polarization is parallel to the quantization axis ($\theta = 0$)

4.1.3.5 556 nm Lasers for MOT

We have two 556 nm laser systems for the MOT of ^{171}Yb and ^{174}Yb. The first laser is consisted of a commercial 1112 nm laser (Toptica, TA pro), an additional homemade TA, and two PPLN-based SHG modules (NTT electronics). The output power of the 556 nm is approximately 200 mW for both modules. The first module is used for the MOT of ^{171}Yb and the second is for Z gate operation on the nuclear spin qubits. The fundamental light is frequency stabilized with a ULE cavity, which is the same as for the 507 nm laser and has a finesse of 2×10^4 for this wavelength, resulting in a laser linewidth of approximately 50 kHz.

The second laser system is consisted of a commercial 1112 nm laser (Toptica, TA pro), and a PPLN-based SHG module (NTT electronics), and used for the MOT of ^{174}Yb. The output power of the 556 nm laser is approximately 200 mW. The frequency of the fundamental light is stabilized by a offset lock to the first 1112 nm laser descrived above.

4.1.3.6 556 nm Laser for Nuclear Spin Manipulation

The 556 nm light for the nuclear spin manipulation is generated by a commercial 1112 nm fiber laser (Menlo systems, Orange-one) and two PPLN-based SHG modules (NTT electronics). The output power of the 556 nm laser is approximately 200 mW for both modules. The formar is used for Raman sideband cooling and the latter is for X gate operation on the nuclear spin qubits. The frequency of the fundamental light is stabilized by a wavemeter.

4.1.3.7 649 and 770 nm Lasers for Repumping

The 649 and 770 nm lasers for repumping of the metastable $^3P_2$ state are IFDLs (Optoquest) with a power of approximately 10 and 20 mW, respectively. The frequency of the lasers are stabilized by a wavemeter.

4.1.3.8 325 nm Laser for Rydberg Excitation

The 325 nm laser for Rydberg excitation is consisted of a 1300 nm laser (Toptica, TA pro), a Raman fiber amplifier (MPB communications) and waveguide SHG module, and a homemade SHG cavity with a BBO crystal. The details of this laser system are described in Chap. 7.

4.1.4 Objective Lenses

We have two custom-made objective lens made by Special Optics. One is for the optical tweezer array generation and 556 nm fluorescence imaging of single atoms, and the other is for the 399 nm fluorescence imaging. Hereafter, the formar is referred as the visible (VIS) objective and the latter is the ultraviolet (UV) objective. The specifications of the objective lenses are summarized in Table 4.3. While the VIS objective is designed to have diffraction limited performance for 532, 556, and 759 nm light, the UV objective is designed for 399 and 325 nm.

The VIS objective is mounted on a assembled stage of 3-axis translation stage (Suruga, B71-80-C) and 5-axis stage (Newport, 9082-V-M), as shown in Fig. 4.4a. This stage allows us to adjust the position of the objective lens with a high precision. The UV objective is mounted on a piezo linear Z stage (PI, P-528.ZCD), which is fixed to a hexapod (PI, H811.I2), as shown in Fig. 4.4b. The hexapod is used for fine adjustment of the UV objective position and the piezo stage is for dynamical adjustment of the focus during three-dimensional optical tweezer array experiment.

Table 4.3 Specifications of the objective lenses

	NA	Working distance (mm)	Focal length (mm)	Field of view (μm)
VIS objective	0.6	20.25	20.0	200 × 200
UV objective	0.6	23.38	28.7	200 × 200

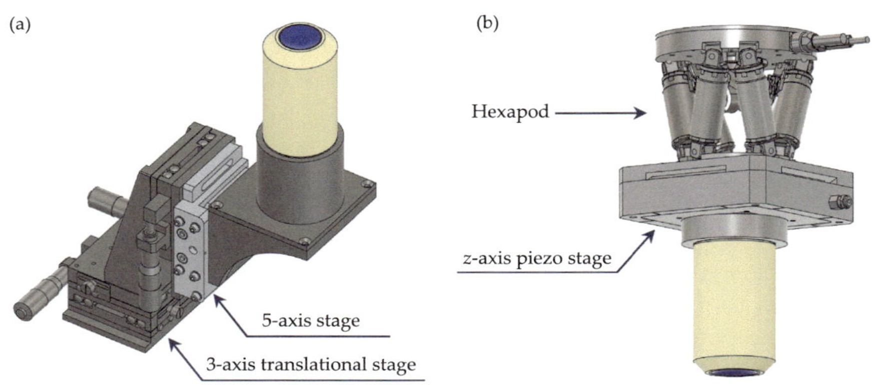

Fig. 4.4 Assembly of the objective and stages. **a** VIS objective mounted on the 3-axis and 5-axis stages. **b** UV objective mounted on the piezo Z stage and the hexapod

4.1.5 Optical Setup

The optical path around the science chamber is shown in Figs. 4.5 and 4.6. On the first floor of the optical table, a diode-pumped solid-state laser (Coherent, Verdi V-10) is placed for optical tweezer array generation. The laser output is firstly divided into two paths: one is for static tweezer array, and the other for the moving tweezers. Each beam is then passes through an acousto-optic modulator (AOM) and coupled to a single mode fiber to clean the beam profile and stabilize the beam pointing. The AOMs are used for switching and power stabilizing of the tweezer beams. The fiber output for static tweezers is delivered to a spatial light modulator (SLM) (Hamamatsu Photonics, X15213-16L). Another fiber output for moving tweezer is sent to an acousto-optic deflector (AOD) (AA Opto-Electronic, DTSX-400-532) for horizontal scanning and then to another AOD for vertical scanning via a $4f$-system. The beams coming from the SLM and AODs are combined by a polarizing beam splitter (PBS) and sent to the second floor of the optical table.

The fiber outputs for MOT Z beam of ^{171}Yb and ^{174}Yb are also placed on the first floor. The beams are combined by a PBS and sent to the second floor. On the second floor, the MOT beams are focused onto the mini mirror, which is a two inches fused silica substrate with a 10 mm mirror coating at the center. This mini mirror configuration allows us to separate the retroreflected MOT Z beam and 556 nm atom fluorescence [4], and thus enables 556 nm fluorescence imaging of the single atoms. The MOT beams reflected by the mini mirror are expanded by a telescope and combined with the optical tweezer beams by a dichroic mirror. Then the combined beams are sent to the science chamber through the VIS objective on the second floor. Note that the MOT beams are focused around the back focal plane of the objective to avoid focusing at the atom position.

The main optical paths, such as MOT XY beams, probe beam, optical pumping (OP) beam, Raman sideband cooling (RSC) beams, metastable excitation beam, repumping beam, and Rydberg excitation beam, are placed on the third floor as shown in Fig. 4.6.

On the fourth and fifth floor, the optical paths for MOT Z beams and atom imaging are placed. The hexapod, which has the piezo Z stage and the UV objective, is inversely fixed to the fifth floor. The MOT Z beams, tweezer beams, and 399 nm atom fluorescence passing through the UV objective are separated by dichroic mirrors on the fifth floor. The MOT Z beams then passes a lens and are retroreflected by a mirror. The atom fluorescence is focused onto an electron-multiplying charge-coupled device (EMCCD) camera (Andor, iXon Ultra 897 EXF) on the fourth floor.

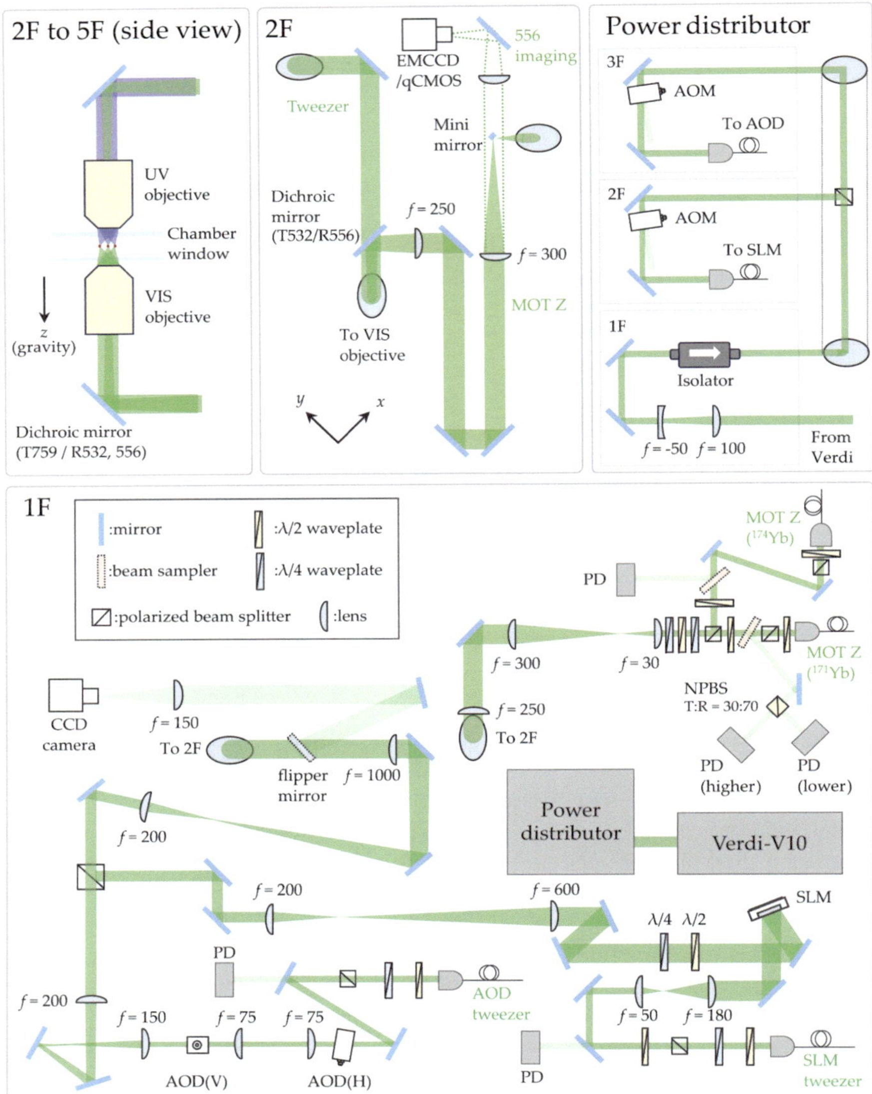

Fig. 4.5 Optical setup of the first and second floor

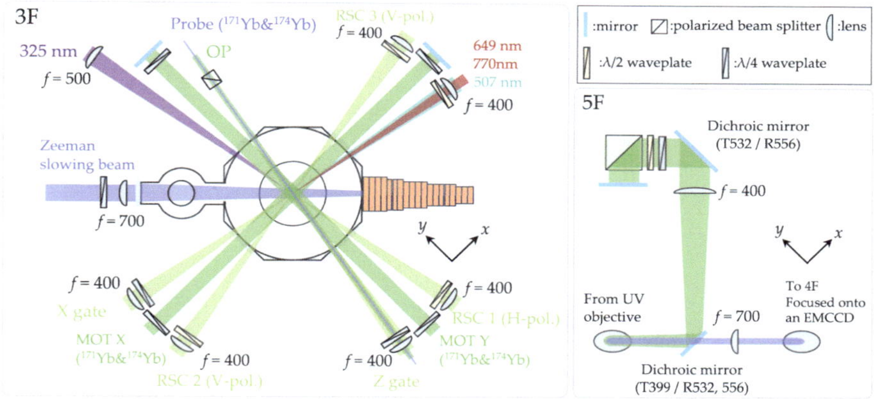

Fig. 4.6 Optical setup of the third and fifth floor

4.2 Experimental Techniques

4.2.1 *Optical Tweezer Array Generation*

The optical tweezer array is the essential technique for this study. In the following, we overview the generation of the optical tweezer array by SLM and AOD.

4.2.1.1 Static Tweezer Array by SLM

The static tweezer array is generated by an SLM. The SLM displays a hologram, which can modulate the wavefront of the incident beam arbitrarily, resulting in arbitrary intensity pattern at the focal plane [5]. We use the weighted Gerchberg-Saxton algorithm [6, 7] to design the hologram for the optical tweezer array.

Let us consider that the incident beam of a wavelength λ has an electric field amplitude A_0, and the target intensity pattern at the focus of a lens with a focal length of f is given as I_t (Fig. 4.7). We iteratively update the phase hologram φ_n on the SLM from a random phase φ_0 to minimize the error between the target intensity pattern and the actual intensity pattern I_n at the focus. We introduce the complex amplitude of the electric field of the k-th trap $V_n^{(k)}$ at the focal plane in the n-th iteration as [7]

$$V_n^{(k)} = \iint A_0 \exp\left[i\left(\varphi_{n-1}(x_s, y_s) - \Delta^{(k)}(x_s, y_s)\right)\right] \mathrm{d}x_s \mathrm{d}y_s, \tag{4.1}$$

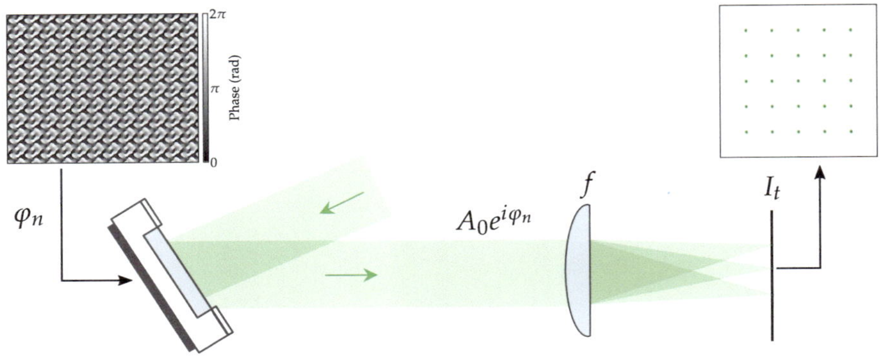

Fig. 4.7 Optical tweezer array generation by SLM

with the pixel coordinates of the SLM (x_s, y_s) and the phase $\Delta^{(k)}$ given as

$$\Delta^{(k)} = \frac{2\pi M}{f\lambda}\left(x^{(k)}x_s + y^{(k)}y_s\right), \tag{4.2}$$

where M is the magnification of the optical system, $\left(x^{(k)}, y^{(k)}\right)$ are the coordinates of the k-th trap. The intensity of the k-th trap $I_n^{(k)}$ is calculated as $I_n^{(k)} = \left|V_n^{(k)}\right|^2$. If the calculated intensity is converged to the target intensity, the iteration is stopped. If not, the phase hologram is updated by the following equations:

$$w_n^{(k)} = w_{n-1}^{(k)}\sqrt{\frac{\langle I_n\rangle}{I_n^{(k)}}}, \tag{4.3}$$

$$\theta_n^{(k)} = \arg\left[V_n^{(k)}\right], \tag{4.4}$$

$$\varphi_n = \arg\left[\sum_k w_n^{(k)}\exp\left[i\left(\Delta^{(k)}(x_s, y_s) + \theta_n^{(k)}\right)\right]\right], \tag{4.5}$$

where $w_n^{(k)}$ is the weight for the site k in the n-th iteration, $\langle I_n\rangle$ is the average of each site intensity $I_n^{(k)}$. This algorithm ensures that the efficiency of the diffraction $e = \sum_k I^{(k)}/\langle I\rangle$ is maximized, and the inhomogeneity of the intensity pattern $\sigma = \sqrt{\langle I^2\rangle - \langle I\rangle^2}/\langle I\rangle$ is minimized (Fig. 4.8a).

Although the algorithm works efficiently and predicts the intensity inhomogeneity of 1%, the actual intensity inhomogeneity at the atom position is around 10% due to the imperfection of the optical system. Therefore, we need to equalize the intensity of the optical tweezers by feedback method. For this, we measure the differential light shift (DLS) of the $^1S_0$ and the $^3P_1$ states in each site using $\left|^1S_0\right\rangle \leftrightarrow \left|^3P_1, m_J = \pm 1\right\rangle$ transition for ^{174}Yb or, $\left|^1S_0\right\rangle \leftrightarrow \left|^3P_1, F = 3/2, m_F = \pm 1/2\right\rangle$ transition for ^{171}Yb. Based on the measured DLS, which is proportional to the intensity of the optical

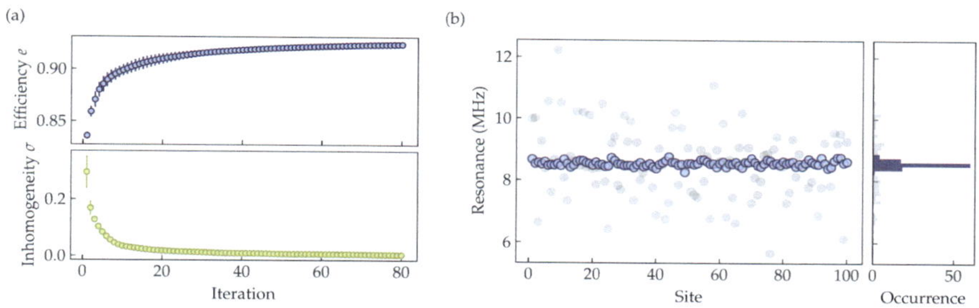

Fig. 4.8 Performance of the WGS algorithm. **a** The calculated diffraction efficiency and inhomogeneity of the tweezer array as a function of the iteration of the WGS algorithm. After 80 iterations, the algorithm converges to the maximum efficiency of 92 % and minimum inhomogeneity of 0.3%. **b** The actual inhomogeneity of the optical tweezer array before and after the equalization via spectroscopy of the $|^1S_0\rangle \leftrightarrow |^3P_1, m_J = \pm 1\rangle$ transition of ^{174}Yb. The intensity inhomogeneity is reduced from 13.3% (gray) to 1.03% (blue)

tweezer, we adjust the intensity of the optical tweezer according to Eq. 4.3. After this feedback, we typically achieve the intensity inhomogeneity of approximately 1% as shown in Fig. 4.8b.

The attractive feature of the SLM is that we can compensate the abberation by adding the Zernike polynomial phases to the hologram. The Zernike polynomial Z_n^m is given as

$$Z_n^m(\rho, \theta) = \begin{cases} R_n^m(\rho)\cos(m\theta) & (m > 0) \\ R_n^{|m|}(\rho)\sin(|m|\theta) & (m < 0), \end{cases} \tag{4.6}$$

with

$$R_n^m(\rho) = \sum_{k=0}^{\frac{n-m}{2}} \frac{(-1)^k (n-k)!}{k!(\frac{n+m}{2} - k)!(\frac{n-m}{2} - k)!} \rho^{n-2k}. \tag{4.7}$$

Each Zernike polynomial corresponds to the specific abberation, such as astigmatism, coma, trefoil, and so on. We optimize the weights a_n^m of the Zernike polynomials to minimize the trap waist, which can be estimated from the trap frequencies measured by Raman sideband spectroscopy of ^{171}Yb, as

$$\varphi_{\text{aber}} = \sum_{m,n} a_n^m Z_n^m(\rho, \theta). \tag{4.8}$$

Figure 4.9 presents the optimized φ_{aber} in our setup.

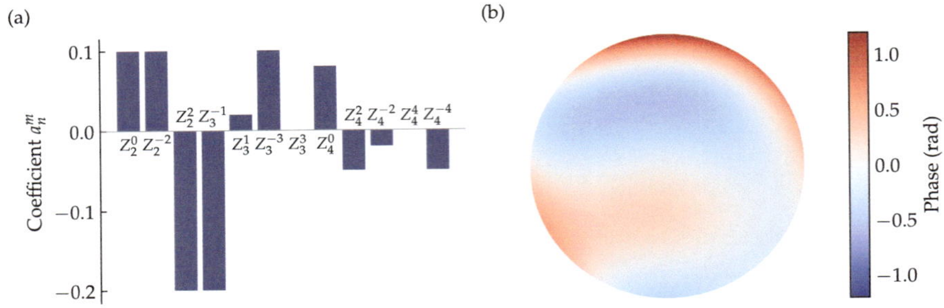

Fig. 4.9 Optimized Zernike polynomial phase for the optical tweezer array in our setup. **a** The coefficients a_n^m for the optimized φ_{aber}. **b** Calculated Zerinike polynomial phase

4.2.1.2 Moving Tweezers by AOD

Moving tweezers play a pivotal role in neutral atom based quantum computer. They are used for making a defect-free atom array [8–10] and for implementing zoned-architecture [11, 12].

The AOD has a crystal inside, which can deflect the incident beam by the Bragg diffraction when radio frequency (RF) signal is applied. The diffraction angle θ_D and the diffracted light frequency f_D are given as

$$\begin{aligned} \theta_D &= \sin^{-1}\left(\frac{c f_{\text{RF}}}{v f_{\text{photon}}}\right) \\ f_D &= f_{\text{photon}} + f_{\text{RF}}, \end{aligned} \tag{4.9}$$

where c is the speed of light, v is the acoustic velocity in the crystal, f_{RF} is the RF frequency, and f_{photon} is the photon frequency. Since the diffraction angle is proportional to the RF frequency, when a multitone RF signal given as

$$V(t) = \sum_{i=1}^{N} A_i \sin(2\pi f_i t + \phi_i), \tag{4.10}$$

is applied to the AOD, N diffracted beams are generated (Fig. 4.10). Here, A_i, f_i, and ϕ_i are the amplitude, frequency, and phase of the i-th RF signal, respectively.

Although the phases $\{\phi_i\}$ seems to be arbitrary, they are important to maintain the diffraction efficiency when the number of the frequency components is large. The limited RF voltage range of the AOD requires low peak-to-average power ratio (PAPR) defined as $V_{\text{max}}^2/\langle V^2\rangle$. If the maximum voltage V_{max} is linearly proportional to the number of frequency components, the amplitudes $\{A_i\}$ have to be reduced not to exceed the applicable voltage range, resulting in low diffraction efficiency. To

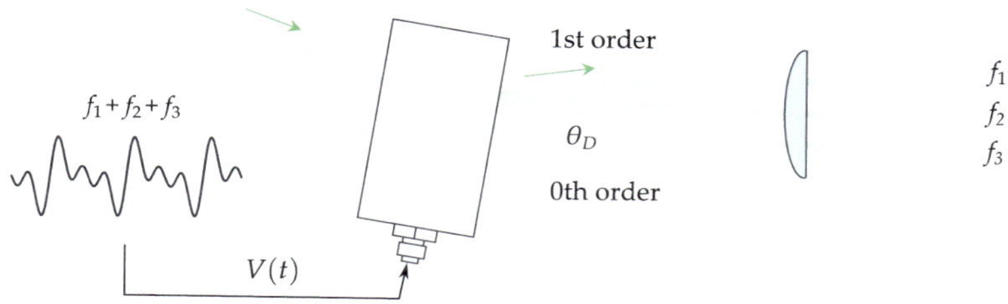

Fig. 4.10 Optical tweezer generation by AOD

avoid this, it is necessary to choose the phases $\{\phi_i\}$ to achieve the PAPR. For this, we employ $\{\phi_i\}$ by Kitayoshi algorithm [13] as

$$\phi_k = \frac{\pi k(k+1)}{N}. \tag{4.11}$$

Figure 4.11 shows the multitone signals with and without employing the Kitayoshi algorithm. The PAPR is reduced from 33.4 to 3.62 for the 16-tone signal.

Frequency sweep of the RF signal can be used for the fast scanning of the optical tweezer, that is moving tweezer. The sweep schedule have to be carefully designed to avoid the atom heating during the transport. We use the following trajectory for the moving tweezer:

$$f(t) = f_s + (f_e - f_s)\sin^2\left(\frac{\pi t}{2T}\right), \tag{4.12}$$

where f_s, f_e are the start and end frequency, and T is the sweep time. This trajectory ensures that the acceleration of the tweezer is zero at the start and end of the sweep. Other groups use different trajectory called minimal jerk trajectory given as

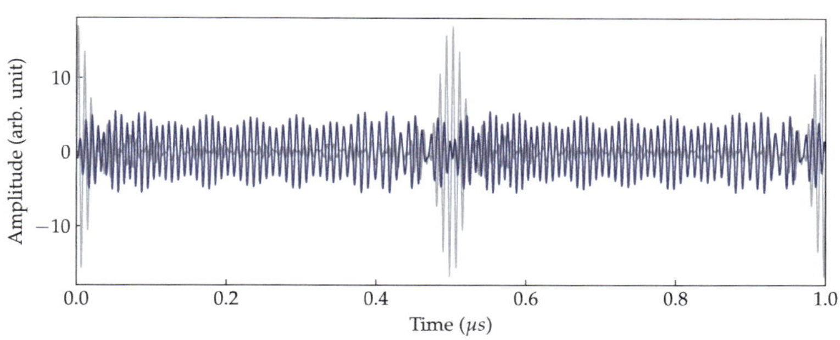

Fig. 4.11 Multitione signal with the Kitayoshi phase, respectively. Gray and blue lines represent 16-tone signals with and without the Kitayoshi phase. The frequencies are equally distributed from 100 to 132 MHz. The PAPR is reduced from 33.4 to 3.62 for this case

$$f(t) = f_s + (f_e - f_s)\left[6\left(\frac{t}{T}\right)^5 - 15\left(\frac{t}{T}\right)^4 + 10\left(\frac{t}{T}\right)^3\right], \tag{4.13}$$

which make the jerk minimal during the transport [11, 14]. Recently, the shortcut to adiabaticity theory is applied to the moving tweezer to minimize the transport time and the atom heating [15].

4.2.2 Light-Assisted Collision

When atoms are loaded into optical tweezers from the MOT, each site initially contains multiple atoms. To prepare single atoms trapped in the tweezers, we utilize light-assisted collisions (LAC). This process induces two-body loss, leading to single-atom occupancy in tweezers with an odd number of atoms. Consequently, the probability of achieving single-atom occupancy is typically 50%.

The LAC process is illustrated in Fig. 4.12. Consider two atoms separated by an interatomic distance R. When both atoms are in the ground state $|S\rangle$, the interatomic potential is described by $V_{|S+S\rangle}(R) = -C_6/R^6$, where C_6 represents the van der Waals interaction coefficient. In contrast, if one atom is in the excited state $|P\rangle$, the potential is given by $V_{|S+P\rangle}(R) = \hbar\omega_0 \pm C_3/R^3$, where ω_0 is the resonant frequency of the $|S\rangle \leftrightarrow |P\rangle$ transition, and C_3 is the dipole-dipole interaction coefficient. When a red-detuned light with frequency ω is applied and the interatomic distance R_c satisfies the condition $V_{|S+P\rangle}(R_c) - V_{|S+S\rangle}(R_c) = \hbar\omega$, one atom is excited to the $|P\rangle$ state.

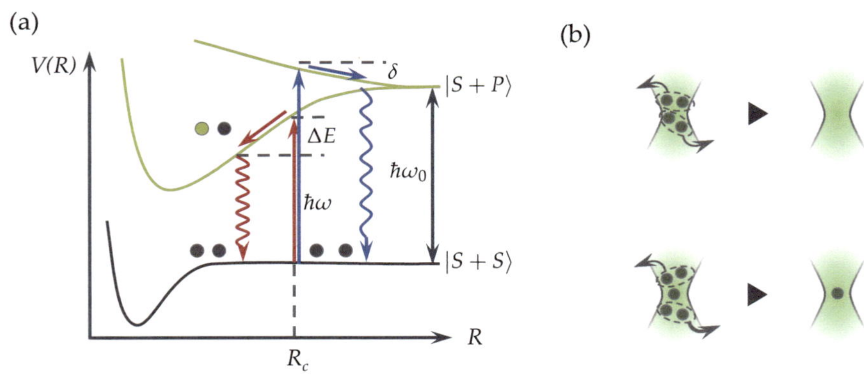

Fig. 4.12 Light-assisted collision process. **a** When atoms are excited by a red-detuned LAC beam, they experience an attractive potential, resulting in the acquisition of kinetic energy ΔE as the interatomic distance decreases. Conversely, excitation with a blue-detuned LAC beam allows control of the kinetic energy up to the detuning δ, thereby enhancing the loading probability. **b** After the LAC process, no atoms remain in the tweezers initially containing an even number of atoms, whereas a single atom persists in the tweezers with an odd number of atoms. For red-detuned LAC, the single-atom loading probability is approximately 50%

Due to the attractive nature of the interatomic potential in this configuration, the atoms acquire kinetic energy as they move closer together. After the excited atom decays back to the ground state, the atoms gain an energy ΔE, which is sufficient for them to escape the optical tweezer potential. This results in two-body loss.

If the trap volume is so small that the loading rate of atoms is much smaller than the two-body loss rate, the number of atoms in the tweezer after LAC is restricted to zero or one. This phenomenon is called collisional blockade [16].

In the above discussion of LAC, we assumed the red-detuned LAC beam. However, the blue-detuned LAC beam can also be used to induce LAC. Notablly, this process enhance the single atom loading probability up to 90% [4, 10, 17–20]. In this case, the blue-detuned LAC beam with a frequency of $\omega_0 + \delta$ excite one of the atoms to the $|P\rangle$ state. The interatomic potential is then repulsive, and the atoms acquire kinetic energy up to δ. When the detuning δ is larger than the trap depth, but smaller than twice of the trap depth, only one atoms can escape from the trap. Consequently, the single atom loading probability is enhanced.

In our setup, we observed the enhanced loading by the blue-detuned LAC transition in the ^{171}Yb tweezer array. We use the $\left|^1\mathrm{S}_0\right\rangle \leftrightarrow \left|^3\mathrm{P}_1, F = 3/2\right\rangle$ transition for this purpose. Figure 4.13 shows the loading probability as a function of the magnetic field and LAC beam detuning. When the magnetic field along the tweezer polarization is set to 2.5 G, the loading probability is increased up to 78%. Further enhancement can be achieved by increasing the atomic density around the tweezer during the loading process.

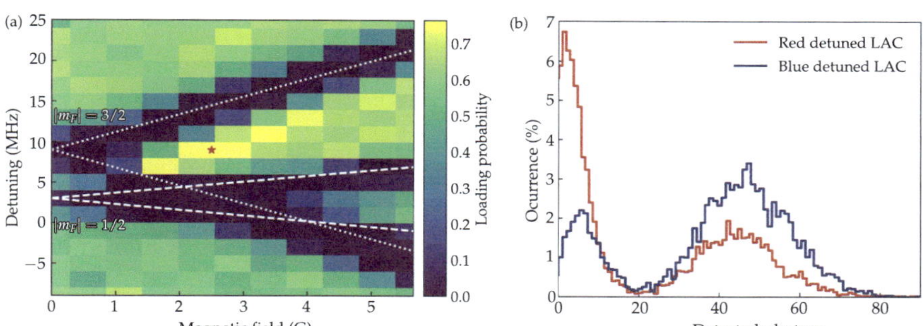

Fig. 4.13 Blue-detuned LAC. **a** Loading probability as a function of the magnetic field and LAC beam detuning. When the magnetic field and the LAC detuning are set to around 2.5 G and 9.5 MHz from the free space resonance, respectively, the loading probability is enhanced. The white dashed and dotted line represent the resonances of $|m_F| = 1/2$ and $|m_F| = 3/2$ states, respectively. **b** Histograms of detected photons. When we set the LAC parameters to the optimal point, which is indicated by a star in (**a**), the loading probability is increased from 45 to 78%

4.2.3 Single Atom Imaging

The single atom imaging is performed by the fluorescence imaging technique. In this study, we use the 399 nm beam with an intensity of $1 \times 10^{-3} I_{\mathrm{sat},399}$ as a probe, and the 556 nm beams, which are the same as the MOT beams, with a total intensity of $10 I_{\mathrm{sat},556}$ as a simultaneous cooling light. Note that the 556 nm fluorescence imaging is also possible in our setup [21], but the 532 nm tweezer beam induces autofluorescence around 556 nm at the objective housing [22], preventing high fidelity imaging in large scale array.

The photon emission pattern from the atoms obeys the dipole radiation pattern $\Phi(\theta, \phi)$, which is given as [23]

$$\Phi(\theta, \phi) = \begin{cases} \dfrac{3}{8\pi} \sin^2 \theta & (\Delta m = 0) \\ \dfrac{3}{16\pi}(1 + \cos^2 \theta) & (\Delta m = \pm 1), \end{cases} \tag{4.14}$$

where θ is the angle from the quantization axis and Δm is the change of the magnetic sublevels in the transition. To maximize the photon collection efficiency, we set the probe beam polarization to be linear and in xy-plane, that is orthogonal to the objective axis (z-axis). When the quantization axis is set to be the z-axis, the photon collection efficiency η is given as

$$\begin{aligned} \eta &= \int_0^{2\pi} d\phi \int_0^{\alpha} d\theta \sin\theta \Phi(\theta, \phi) = \frac{3}{16\pi} \int_0^{2\pi} d\phi \int_0^{\alpha} d\theta \sin\theta (1 + \cos^2 \theta) \\ &= \frac{1}{8} \left[4 - \cos\alpha (3 + \cos^2 \alpha)\right] \\ &= \frac{1}{8} \left[4 - (4 - \mathrm{NA}^2)\sqrt{1 - \mathrm{NA}^2}\right], \end{aligned} \tag{4.15}$$

where $\mathrm{NA} = \sin\alpha$ is the numerical aperture of the objective. Similarly, when the quantization axis is set to be in the xy-plane, the photon collection efficiency is given as

$$\begin{aligned} \eta &= \int_0^{2\pi} d\tilde{\phi} \int_0^{\alpha} d\tilde{\theta} \sin\tilde{\theta} \Phi(\tilde{\theta}, \tilde{\phi}) = \frac{3}{8\pi} \int_0^{2\pi} d\tilde{\phi} \int_0^{\alpha} d\tilde{\theta} \sin\tilde{\theta} (\cos^2 \tilde{\theta} + \sin^2 \tilde{\theta} \cos^2 \tilde{\phi}) \\ &= \frac{1}{8} \left[4 - \cos\alpha (3 + \cos^2 \alpha)\right] \\ &= \frac{1}{8} \left[4 - (4 - \mathrm{NA}^2)\sqrt{1 - \mathrm{NA}^2}\right]. \end{aligned} \tag{4.16}$$

Here we take a rotated frame $(\tilde{\theta}, \tilde{\phi})$ so that the objective axis corresponds to $\tilde{\theta} = 0$ axis. In this frame, $\sin^2 \theta = \cos^2 \tilde{\theta} + \sin^2 \tilde{\theta} \cos^2 \tilde{\phi}$. From the above calculations, we find that the above two cases have the same photon collection efficiency.

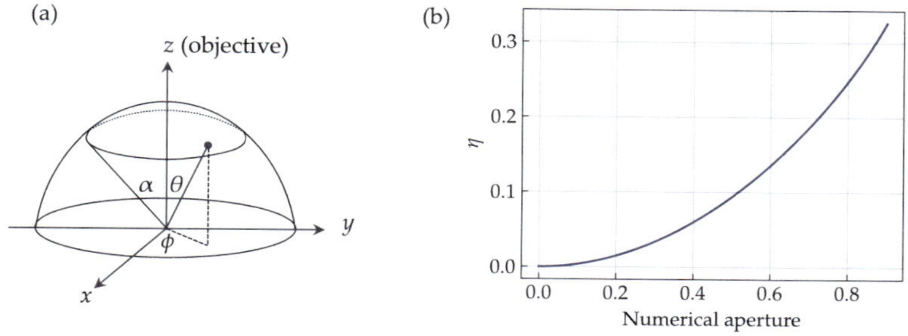

Fig. 4.14 Calculation of the photon collection efficiency. **a** Coordinates for the calculation. **b** Photon collection efficiency as a function of the numerical aperture of the objective

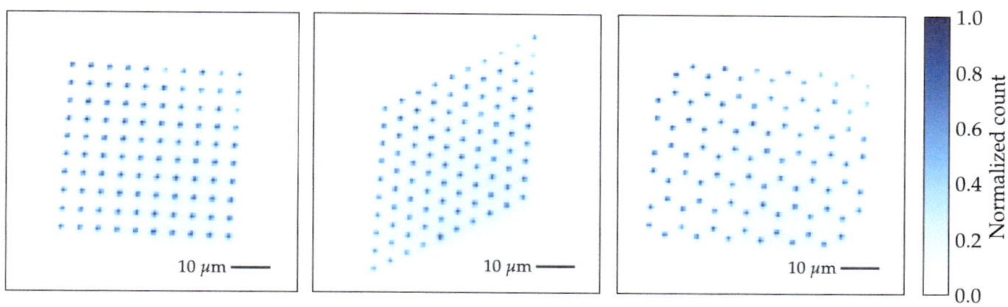

Fig. 4.15 Averaged fluorescence images of single ^{174}Yb atoms trapped in various tweezer arrays. (left) Square array of 600 shots average. (middle) Triangle array of 300 shots average. (right) Honeycomb array of 300 shots average

Figure 4.14 shows the photon collection efficiency as a function of the numerical aperture of the objective. In our setup, the NA of the objective is 0.6, corresponding to the theoretical photon collection efficiency of 13.6%.

Figure 4.15 presents the 399 nm fluorescence image of single ^{174}Yb atoms trapped in various tweezer arrays. The imaging performance will be discussed in Sect. 6.3.1.1.

4.2.4 Release and Recapture of Single Atoms

The temperature of the single atoms in the optical tweezer array can be estimated by release and recapture method [24]. In this method, the atoms are firstly released from the optical tweezer array by turning off the optical tweezer beams. After certain time of flight Δt, the tweezer beams are turned on again to recapture the atoms. If the temperature of the atoms is high, the atoms are spread out and the recapture probability is rapidly reduced. By measuring the recapture probability as a function of Δt, we can estimate the temperature of the atoms by comparing the experimental data with the Monte Carlo simulation [25].

The procedure of the Monte Carlo simulation is as follows: (1) Generate the initial position (x_i, y_i, z_i) and velocity (v_{xi}, v_{yi}, v_{zi}) of the atoms. The initial position is normally distributed with a standard deviation of $\sigma_{x,y,z} = \sqrt{k_B T/(m\omega^2_{x,y,z})}$ as

$$n(x, y, z) = \frac{1}{(2\pi)^{3/2}\sigma_x\sigma_y\sigma_z}\exp\left[-\left(\frac{x^2}{2\sigma_x^2} - \frac{y^2}{2\sigma_y^2} - \frac{z^2}{2\sigma_z^2}\right)\right], \tag{4.17}$$

and the initial velocity obeys the Maxwell-Boltzmann distribution given as

$$f(v) = 4\pi v^2 \left(\frac{m}{2\pi k_B T})^{3/2}\exp(-\frac{mv^2}{2k_B T}\right). \tag{4.18}$$

Here, m is the mass of the atom, T is the atom temperature, k_B is the Boltzmann constant, and $\omega_{x,y,z}$ are the trap frequencies along each direction. Once the velocity is determined, the direction of the velocity is chosen from the uniform distribution in the solid angle. (2) Calculate the atom position $\boldsymbol{r}$ and velocity $\boldsymbol{v}$ after time of flight Δt as $\boldsymbol{r} = (x_i + v_{xi}\Delta t, y_i + v_{yi}\Delta t, z_i + v_{zi}\Delta t - g\Delta t^2/2)$ and $\boldsymbol{v} = (v_{xi}, v_{yi}, v_{zi} - gt)$, where g is the gravitational acceleration. (3) Check if the kinetic energy of the atom $m\boldsymbol{v}^2/2$ is larger than the effective trapping potential at $\boldsymbol{r}$, resulting in the atom loss. Note that the trapping potential includes the gravitational potential. (4) Repeat the above steps for N samples and calculate the recapture probability. (5) Fit the experimental data with the Monte Calro simulation by least square method. The fit parameters are the temperature of the atoms and the survival probability at $\Delta t = 0$.

Figure 4.16 shows the release and recapture experiment of ^{174}Yb. The temperature of the atoms is estimated to be 35(3)μK, which is higher than the Doppler cooling limit of 4.4 μK, and can be reduced by optimizing the intensity balance of the cooling beams. The mismatch of the experimental data and the Monte Carlo simulation at longer release time might be attributed to the irregularity of the tweezer beam shape due to the abberations, which is not considered in the simulation.

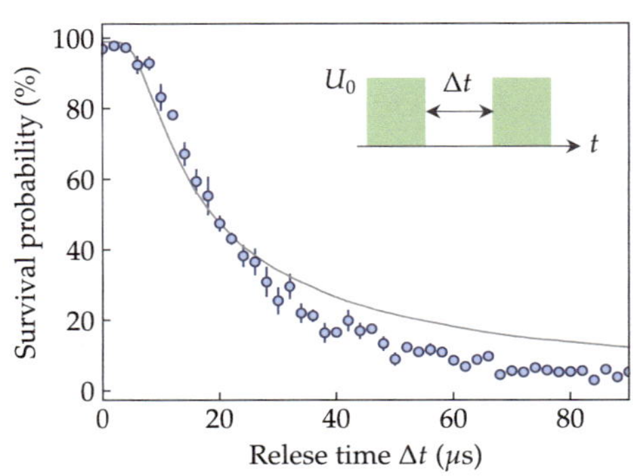

Fig. 4.16 Release and recapture experiment of ^{174}Yb. The recapture probability as a function of the time of flight Δt. The gray line is the fitting result from 100 trials of the Monte Carlo simulation. The temperature is estimated to be 35(3)μK

References

1. D. Okuno, Single ytterbium atoms in an optical tweezer array: high-resolution spectroscopy, single-photon Rydberg excitation, and a scheme for nondestructive detection. Ph.D. Thesis, Kyoto University (2022). https://repository.kulib.kyoto-u.ac.jp/dspace/bitstream/2433/276155/1/drigk04851.pdf
2. R. Yamamoto, J. Kobayashi, T. Kuno, K. Kato, Y. Takahashi, New J. Phys. **18**(2), 023016 (2016). https://doi.org/10.1088/1367-2630/18/2/023016. https://iopscience.iop.org/article/10.1088/1367-2630/18/2/023016/meta
3. S. Kato, High-resolution laser spectroscopy of quantum degenerate ytterbium atoms in an optical lattice. Ph.D. Thesis, Kyoto University (2012). https://repository.kulib.kyoto-u.ac.jp/dspace/handle/2433/161035
4. A. Jenkins, J.W. Lis, A. Senoo, W.F. McGrew, A.M. Kaufman, Phys. Rev. X **12**(2), 021027 (2022). https://doi.org/10.1103/PhysRevX.12.021027. https://link.aps.org/doi/10.1103/PhysRevX.12.021027
5. F. Nogrette, H. Labuhn, S. Ravets, D. Barredo, L. Béguin, A. Vernier, T. Lahaye, A. Browaeys, Phys. Rev. X **4**(2), 021034 (2014). https://doi.org/10.1103/physrevx.4.021034. https://link.aps.org/doi/10.1103/PhysRevX.4.021034
6. R. Di Leonardo, F. Ianni, G. Ruocco, Opt. Express **15**(4), 1913 (2007). https://doi.org/10.1364/oe.15.001913
7. S. de Léséleuc, Quantum simulation of spin models with assembled arrays of Rydberg atoms. Ph.D. Thesis, Université Paris Saclay (2018). https://hal-obspm.ccsd.cnrs.fr/STAR/tel-02088297v1
8. M. Endres, H. Bernien, A. Keesling, H. Levine, E.R. Anschuetz, A. Krajenbrink, C. Senko, V. Vuletic, M. Greiner, M.D. Lukin, Science **354**(6315), 1024 (2016). https://doi.org/10.1126/science.aah3752. https://www.science.org/doi/abs/10.1126/science.aah3752
9. D. Barredo, S. de Léséleuc, V. Lienhard, T. Lahaye, A. Browaeys, Science **354**(6315), 1021 (2016). https://doi.org/10.1126/science.aah3778
10. W. Tian, W.J. Wee, A. Qu, B.J.M. Lim, P.R. Datla, V.P.W. Koh, H. Loh, Phys. Rev. Appl. **19**(3), 034048 (2023). https://doi.org/10.1103/physrevapplied.19.034048. https://link.aps.org/doi/10.1103/PhysRevApplied.19.034048
11. D. Bluvstein, H. Levine, G. Semeghini, T.T. Wang, S. Ebadi, M. Kalinowski, A. Keesling, N. Maskara, H. Pichler, M. Greiner, V. Vuletić, M.D. Lukin, Nature (London) **604**(7906), 451 (2022). https://doi.org/10.1038/s41586-022-04592-6. https://www.nature.com/articles/s41586-022-04592-6
12. D. Bluvstein, S.J. Evered, A.A. Geim, S.H. Li, H. Zhou, T. Manovitz, S. Ebadi, M. Cain, M. Kalinowski, D. Hangleiter, J.P. Bonilla Ataides, N. Maskara, I. Cong, X. Gao, P. Sales Rodriguez, T. Karolyshyn, G. Semeghini, M.J. Gullans, M. Greiner, V. Vuletić, M.D. Lukin, Nature (London) **626**(7997), 58 (2024). https://doi.org/10.1038/s41586-023-06927-3
13. H. Kitayoshi, S. Sumieda, K. Sirakawa, S. Takeshita, in *Digest of Papers - International Test Conference* (1985), pp. 825–834. https://tohoku.pure.elsevier.com/en/publications/dsp-synthesized-signal-source-for-analog-testing-stimulus-and-new
14. R. Finkelstein, R.B.S. Tsai, X. Sun, P. Scholl, S. Direkci, T. Gefen, J. Choi, A.L. Shaw, M. Endres, Nature **634**(8033), 321 (2024). https://doi.org/10.1038/s41586-024-08005-8. https://www.nature.com/articles/s41586-024-08005-8
15. S. Hwang, H. Hwang, K. Kim, A. Byun, S. Jeong, M.P. Soegianto, J. Ahn (2014). arXiv:2410.22627
16. N. Schlosser, G. Reymond, P. Grangier, Phys. Rev. Lett. **89**(2), 023005 (2002). https://doi.org/10.1103/PhysRevLett.89.023005
17. T. Grünzweig, A. Hilliard, M. McGovern, M.F. Andersen, Nat. Phys. **6**(12), 951 (2010). https://doi.org/10.1038/nphys1778. https://www.nature.com/articles/nphys1778
18. Y.H. Fung, M.F. Andersen, New J. Phys. **17**(7), 073011 (2015). https://doi.org/10.1088/1367-2630/17/7/073011. https://iopscience.iop.org/article/10.1088/1367-2630/17/7/073011/meta

19. B.J. Lester, N. Luick, A.M. Kaufman, C.M. Reynolds, C.A. Regal, Phys. Rev. Lett. **115**(7), 073003 (2015). https://doi.org/10.1103/PhysRevLett.115.073003
20. M.O. Brown, T. Thiele, C. Kiehl, T.W. Hsu, C.A. Regal, Phys. Rev. X **9**(1), 011057 (2019). https://doi.org/10.1103/PhysRevX.9.011057. https://link.aps.org/doi/10.1103/PhysRevX.9.011057
21. D. Okuno, Y. Nakamura, T. Kusano, Y. Takasu, N. Takei, H. Konishi, Y. Takahashi, J. Phys. Soc. Jpn. **91**(8), 084301 (2022). https://doi.org/10.7566/JPSJ.91.084301
22. S. Saskin, J.T. Wilson, B. Grinkemeyer, J.D. Thompson, Phys. Rev. Lett. **122**(14), 143002 (2019). https://doi.org/10.1103/PhysRevLett.122.143002
23. D.A. Steck, *Quantum and Atom Optics* (2007). https://atomoptics.uoregon.edu/~dsteck/teaching/quantum-optics/
24. G. Reymond, N. Schlosser, I. Protsenko, P. Grangier, Philos. Trans. A Math. Phys. Eng. Sci. **361**(1808), 1527 (2003). https://doi.org/10.1098/rsta.2003.1219
25. C. Tuchendler, A.M. Lance, A. Browaeys, Y.R.P. Sortais, P. Grangier, Phys. Rev. A **78**(3), 033425 (2008). https://doi.org/10.1103/PhysRevA.78.033425. https://link.aps.org/doi/10.1103/PhysRevA.78.033425

Chapter 5
Coherent Manipulations of Nuclear Spin and Optical Clock Qubits in Ytterbium Atoms

This chapter outlines the coherent manipulation of qubits in ytterbium atoms. First, we provide an overview of the fundamental properties of nuclear spin qubits in ^{171}Yb and present experimental results, including initialization, readout, and Raman sideband cooling. Subsequently, we focus on optical clock qubits in ^{174}Yb, highlighting the observation of coherent Rabi oscillations and the Ramsey interference experiment on the optical qubits.

5.1 Nulear Spin Qubits in ^{171}Yb

5.1.1 Basic Properties

The nuclear spin states in the $^1S_0$ and $^3P_0$ states of ^{171}Yb are ideal candidates for qubits [1–6]. This suitability arises from the fact that the total angular momentum of the valence electrons in these states is zero, eliminating hyperfine coupling between the nuclear spin and the electron. Furthermore, the nuclear spin states exhibit remarkable insensitivity to magnetic fields, with a Zeeman splitting of only 751 Hz/G. As a result, nuclear spin qubits demonstrate exceptionally long coherence times [1, 2, 4], typically three orders of magnitude longer than those of hyperfine clock qubits in alkali atoms [7]. In this study, we specifically encode the qubit states using the ground-state nuclear spin states as $|0\rangle = |^1S_0, m_F = 1/2\rangle$ and $|1\rangle = |^1S_0, m_F = -1/2\rangle$.

A particularly compelling feature of ^{171}Yb is its nuclear spin of $1/2$, which stands in contrast to the nuclear spin of $9/2$ in strontium-87 atoms. This minimal nuclear spin structure enables simplified quantum state control and significantly facilitates fast and high-fidelity manipulations of the nuclear spin states via Raman transitions [1], as will be discussed in later sections. These attributes position ^{171}Yb as a promising platform for scalable quantum information processing.

Y. Nakamura, *Generation of Highly Controllable Ytterbium Atom Array*,
Springer Theses, https://doi.org/10.1007/978-981-95-2836-3_5

5.1.2 State Initialization and Readout

The state initialization is achieved through optical pumping (OP) via the $\left|{}^1\mathrm{S}_0, m_F = -1/2\right\rangle \rightarrow \left|{}^3\mathrm{P}_1, F = 3/2, m_F = 1/2\right\rangle$ transition (Fig. 5.1). This process is implemented by irradiating a 556 nm laser beam in the horizontal plane under a 48 G magnetic field applied along the z-axis. The OP beam has a radius of 1 mm, an intensity of $3I_{\mathrm{sat}}$, and is linearly polarized within the horizontal plane.

The state readout is performed by selectively blasting out the $|0\rangle$ state using the $\left|{}^1\mathrm{S}_0, m_F = 1/2\right\rangle \leftrightarrow \left|{}^3\mathrm{P}_1, F = 3/2, m_F = 3/2\right\rangle$ transition (Fig. 5.1), followed by fluorescence imaging as described in Sect. 4.2.3. The blast beam is identical to the OP beam, except that its intensity is increased to $9I_{\mathrm{sat}}$.

The population of the qubit states during the blast obeys following rate equations:

$$\begin{cases} \frac{\mathrm{d}N_0}{\mathrm{d}t} = -LN_0 + PN_1, \\ \frac{\mathrm{d}N_1}{\mathrm{d}t} = -PN_1, \end{cases} \tag{5.1}$$

where N_0 and N_1 is the population of $|0\rangle$ and $|1\rangle$ states, respectively, L is the loss rate of the $|0\rangle$ state, and P is the unintended pumping rate from the $|1\rangle$ state to the $|0\rangle$ state induced by the blast beam. The solution of these equations is given by

$$\begin{cases} N_0(t) = \frac{PN_1(0)}{P-L}(e^{-Lt} - e^{-Pt}) + N_0(0)e^{-Lt}, \\ N_1(t) = N_1(0)e^{-Pt}. \end{cases} \tag{5.2}$$

By measuring the populations of the qubit states without applying OP, we can evaluate the performance of the blasting operation. Figure 5.2a shows the population of the $|0\rangle$ state as a function of blast time. Fitting the data using Eq. (5.2) yields a loss rate of $L = 3.5(3)$ kHz, with the unintended pumping rate suppressed to $P <$ 5 Hz. Under these conditions, blasting for 5 ms removes more than 99.9% of the $|0\rangle$ state while preserving 98(3)% of the $|1\rangle$ state. Further improvement in the blasting performance can be achieved by increasing the strength of the magnetic field during

Fig. 5.1 Relevant energy levels for OP and blast

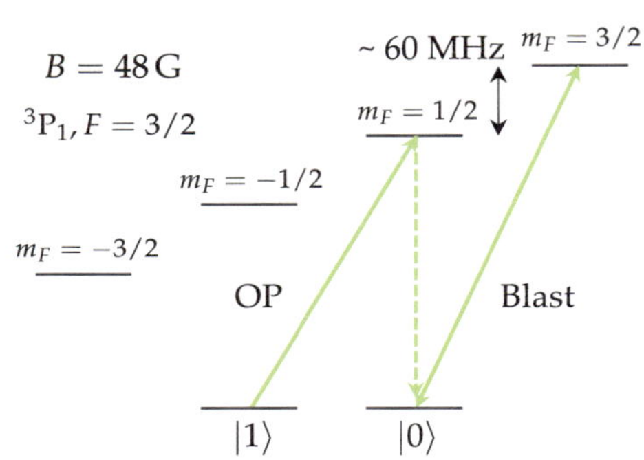

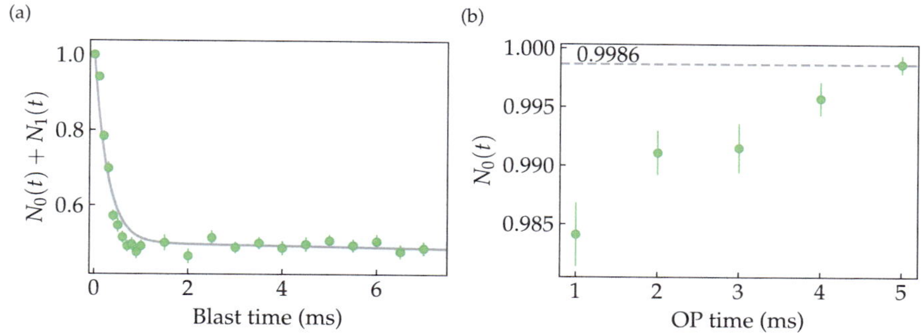

Fig. 5.2 Evaluation of the blast and OP performance. **a** Population dynamics as a function of blast time. The solid line represents the fit to the data using Eq. (5.2). The shaded area represents the 1σ confidence intervals of the fit. From the fitting, the blast rate and unintended pumping rate is estimated to be $L = 3.5(3)$ kHz and $P < 5$ Hz, respectively. Note that $N_0(0) = N_1(0) = 1/2$ is assumed in this calculation. **b** Population of the $|0\rangle$ state as a function of OP time. After applying the OP pulse of 5 ms, 99.86(8)% of the atoms in the $|0\rangle$ state. The error bars in this figure represent the standard error of the mean

the blasting process. A stronger magnetic field enhances the Zeeman splitting of the $^3P_1$ state, effectively suppressing unintended pumping transitions and increasing the selectivity of the operation.

The performance of the OP process is evaluated by measuring the population of the $|0\rangle$ state as a function of OP time. Figure 5.2b illustrates the population of the $|0\rangle$ state after the blasting operation. Assuming that 99.9% of the atoms in the $|0\rangle$ state are removed after the blasting process, the data indicate that 99.86(8)% of the atoms are pumped into the $|0\rangle$ state after an OP time of 5 ms.

5.1.3 Raman Transitions Between the Qubit States

The Raman transitions are essential for the coherent manipulation of nuclear spin qubits and for Raman sideband cooling. Before presenting the experimental results, we provide an overview of the Raman transitions between nuclear spin states.

5.1.3.1 Formulation of the Raman Transition

The strength of the transition, characterized by the Rabi frequency between the hyperfine states $|F_g, m_g\rangle$ and $|F_e, m_e\rangle$, is proportional to the matrix element $\left\langle F_g, m_g \middle| \hat{d}_q \middle| F_e, m_e \right\rangle$. Here, F_g and F_e represent the total angular momentum of the ground and excited states, respectively, while m_g and m_e are their projections onto the quantization axis. The operator $\hat{d}_q$ is the q-component of the electric dipole operator in the spherical basis. According to the Wigner-Eckart theorem, the matrix element can be expressed as

$$\left\langle F_g, m_g \middle| \hat{d}_q \middle| F_e, m_e \right\rangle = \frac{\langle F_g, m_g | \; F_e, m_e; 1, q\rangle}{\sqrt{2F_g + 1}} \langle F_g || \boldsymbol{d} || F_e \rangle, \tag{5.3}$$

where $\boldsymbol{d}$ is the electric dipole operator. The reduced matrix element is further decomposed as

$$\langle F_g || \boldsymbol{d} || F_e \rangle = (-1)^{F_e + J_g + 1 + I} \sqrt{(2F_g + 1)(2F_e + 1)} \begin{Bmatrix} J_g & J_e & 1 \\ F_e & F_g & I \end{Bmatrix} \langle J_g || \boldsymbol{d} || J_e \rangle. \tag{5.4}$$

Here, J_g and J_e denote the total electronic angular momentum of the ground and excited states, respectively, and I is the nuclear spin. The atom-light interaction Hamiltonian for the transition between the hyperfine states can be written as

$$\hat{\mathcal{H}}_{\text{int}} = \frac{\hbar}{2} \sum_q \left(\Omega_q \hat{\Sigma}_q + \Omega_q^* \hat{\Sigma}_q^\dagger \right), \tag{5.5}$$

with the Rabi frequency Ω_q given by

$$\Omega_q = -\frac{\langle J_g || \boldsymbol{d} || J_e \rangle E_q}{\hbar}, \tag{5.6}$$

where E_q is the electric field amplitude of the light, and

$$\hat{\Sigma}_q = \sum_{F_e} \sum_{m_g, m_e} S_{m_g m_e}^{(q)}(F_e) \left| F_g, m_g \right\rangle \langle F_e, m_e | . \tag{5.7}$$

Here, we define the coefficients $S_{m_g m_e}^{(q)}(F_e)$ as

$$S_{m_g m_e}^{(q)}(F_e) := (-1)^{F_e + J_g + 1 + I} \sqrt{2F_e + 1} \left\langle F_g, m_g \right| \; F_e, m_e; 1, q\rangle \begin{Bmatrix} J_g & J_e & 1 \\ F_e & F_g & I \end{Bmatrix}. \tag{5.8}$$

Let us consider a three-level system consisting of nearly-degenerated ground states of $|g\rangle = \left|F_g, m_g\right\rangle$ and $\left|g'\right\rangle = \left|F_g, m_g'\right\rangle$, and an excited state of $|e\rangle = |F_e, m_e\rangle$. The Raman transition between the $|g\rangle$ and $\left|g'\right\rangle$ states is driven by two laser beams with a frequency difference of δ and a detuning of Δ_{F_e} from the $|g\rangle \rightarrow |e\rangle$ and $\left|g'\right\rangle \rightarrow |e\rangle$ transitions. When the frequency difference δ matches the energy difference between the $|g\rangle$ and $\left|g'\right\rangle$ states, and the angular momentum change of the photons $q' - q$ satisfies the conservation law of the angular momentum $q' - q = m_g - m_g'$, the Raman transition occurs. Under these conditions, the effective Rabi frequency of the Raman transition is given by:

$$\Omega^{qq'}_{m_g m'_g} = \Omega_q \Omega^*_{q'} \sum_{F_e} \sum_{m_e} \frac{S^{(q)}_{m_g m_e}(F_e) S^{(q')}_{m'_g m_e}(F_e)}{2\Delta_{F_e}}$$
$$= \frac{3\pi c^2}{\hbar\omega_0^3}(2J_e+1)\Gamma \sum_{F_e} \sum_{m_e} \frac{S^{(q)}_{m_g m_e}(F_e) S^{(q')}_{m'_g m_e}(F_e)}{\Delta_{F_e}} \sqrt{I_q I_{q'}}. \quad (5.9)$$

Here, ω_0 is the resonance frequency of the $|g, g'\rangle \to |e\rangle$ transition, $I_q = \epsilon_0 c E_q^2/2$ is the intensities of the laser beams, and Γ is the natural linewidth of the excited state, which is given by

$$\Gamma = \frac{\omega_0^3}{3\pi\epsilon_0\hbar c^3} \frac{1}{2J_e+1} |\langle J_g||\boldsymbol{d}||J_e\rangle|^2. \quad (5.10)$$

The Raman beams also induce the light shifts and scattering of the ground states. Similar to Eqs. (2.59) and (2.60), the light shift and scattering rate of the $|g\rangle$ state are given by

$$U_{m_g} = \frac{3\pi c^2}{2\omega_0^3}(2J_e+1)\Gamma \sum_q \sum_{F_e} \sum_{m_e} \frac{\left|S^{(q)}_{m_g m_e}(F_e)\right|^2}{\Delta_{F_e,m_e}} I_q, \quad (5.11)$$

$$\gamma_{\mathrm{sc}} = \frac{3\pi c^2}{2\hbar\omega_0^3}(2J_e+1)^2\Gamma^2 \sum_q \sum_{F_e} \sum_{m_e} \frac{\left|S^{(q)}_{m_g m_e}(F_e)\right|^2}{\Delta^2_{F_e,m_e}} I_q. \quad (5.12)$$

Figure 5.3 shows the light shift of the nuclear spin states as a function of the detuning of the Raman beams.

Here, it is worth noting that the above formula for the scattering rate is, strictly speaking, NOT entirely accurate. This is because the formula represents an independent summation of the scattering rates for each transition path and does not account for the detailed scattering processes, such as Raman and Rayleigh scattering. Despite this limitation, the formula provides a practical and convenient approximation for estimating the scattering rate in experimental settings. A more detailed and rigorous discussion on the scattering process, which includes all relevant contributions, will be discussed later.

5.1.3.2 Conventional Raman Transitions

We first demonstrate the coherent manipulation of nuclear spin states using conventional Raman transitions driven by two laser beams. For this purpose, the $^3\mathrm{P}_1$ state is chosen as the intermediate state for the Raman transitions. The beam configuration is illustrated in Figs. 5.4a and 4.6.

Three beams, referred to as RSC1, RSC2, and RSC3, are employed. The RSC1 beam is linearly polarized in the horizontal plane, whereas the RSC2 and RSC3 beams

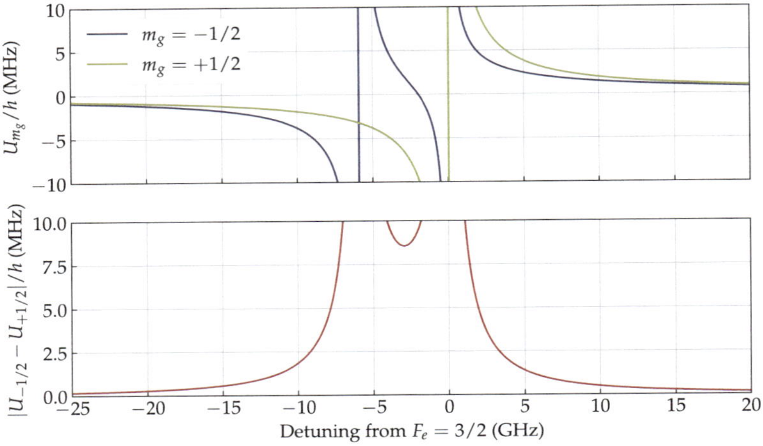

Fig. 5.3 Light shift of the nuclear spin states. (Top) Light shift of each nuclear spin state as a function of the detuning from the $^3P_1$, $F_e = 3/2$ manifolds. (Bottom) Nuclear spin splitting induced by the light shift. In this calculation, we assume that the beam has a power of 100 mW, a beam radius of 100 μm, and circular polarization ($q = 1$)

are linearly polarized in the vertical plane. Each beam has an elliptical intensity profile with a beam radius of approximately 35 μm in the vertical direction and 350 μm in the horizontal direction at the atom position, ensuring uniform intensity across the atom array. The tweezer potential is approximately 1.6 mK deep, and trap frequencies are $2\pi \times (153, 173)$ kHz in the radial directions, respectively.[1]

Under a magnetic field of 48 G applied along the z-axis, the pairs of RSC1 and RSC2, as well as RSC1 and RSC3, are capable of driving Raman transitions between the $|0\rangle$ and $|1\rangle$ states when the frequency difference between the beams matches the Zeeman splitting of the qubit states of 33.8 kHz. Figure 5.4b shows a typical Rabi oscillation between the $|0\rangle$ and $|1\rangle$ states driven by the RSC1 and RSC2 beams with a beam power of 2.5 mW with a detuning of –3 GHz from the $^3P_1$, $F = 3/2$ manifolds. The Rabi frequency is estimated to be $2\pi \times 14.5(1)$ kHz. We also perform coherent manipulation on the blue-sideband Raman transition as shown in blue curve in Fig. 5.4b. The Rabi frequency of the blue-sideband transition is estimated to be $2\pi \times 5.8(3)$ kHz. The low contrast and fast decay of the carrier Rabi oscillation can be attributed to the high atom temperature.

5.1.3.3 Raman Sideband Cooling

As seen in the previous subsection, the atom temperature in the tweezer influences the coherent dynamics of nuclear spin states. In addition to affecting the Raman transi-

[1] The tweezer beam has an elliptical shape with beam radii of 500 and 560 nm along the two principal axes, primarily due to optical aberrations.

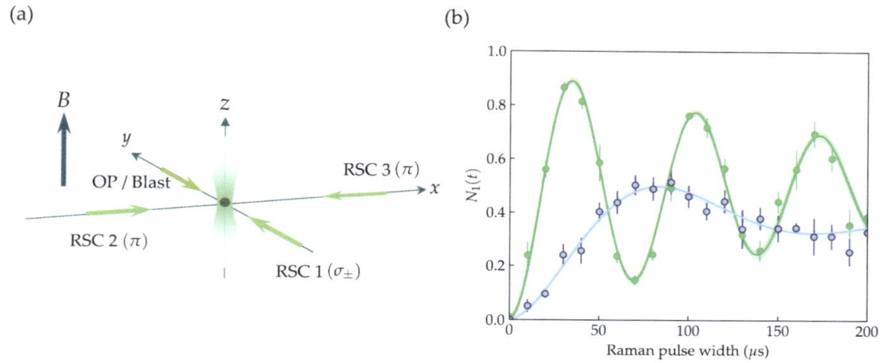

Fig. 5.4 Raman transitions for nuclear spin manipulations. **a** Schematic of the beam configuration for the Raman transitions. **b** Rabi oscillations between the $|0\rangle$ and $|1\rangle$ states, driven by the RSC1 and RSC3 beams. The green circles correspond to the Rabi oscillations on the carrier transition, while the blue circles represent the oscillations on the blue sideband transition. The solid lines are the fits to the data using a damped sinusoidal function. The error bars represent the standard error of the mean

tions, the clock transition–critical for achieving high-fidelity quantum operations–is also impacted by the atom temperature.

To mitigate these effects, we implement Raman sideband cooling (RSC) to cool the atoms to the motional ground state. This section begins with a theoretical discussion of motional dephasing, particularly its impact on the clock transition, to illustrate the necessity of RSC. We then describe the method and present the experimental results of RSC.

The Rabi frequency of a transition between two internal levels of an atom is given by [8]

$$\Omega_n = \Omega_{\text{el}} \prod_{j=x,y,z} \left|\langle n_j | \exp(ik_j j) | n_j \rangle\right|, \tag{5.13}$$

where Ω_{el} is the Rabi frequency of the electronic transition, n_j is the vibrational quantum number along the j-axis, and k_j is the wave number of the laser beam along the j-axis. When the laser propagates along a specific axis, such as the x-axis, the Rabi frequency simplifies to

$$\Omega_n = \Omega_{\text{el}} \exp(-\eta^2/2) L_n(\eta^2). \tag{5.14}$$

Here, $n = n_x$ is the vibratioal quantum number along x-axis, $\eta = k_x\sqrt{\hbar/(2m\omega_x)}$ is the Lamb-Dicke parameter, ω_x is the trap frequency along x-axis, $L_n(\cdot)$ is the n-th order Laguerre polynomial, and we use Eq. (2.94) in this calculation. Consequently, the population dynamics of the excited state becomes

$$\rho_{ee}(t) = \frac{1}{1+\bar{n}} \sum_{n=0}^{\infty} \left(\frac{\bar{n}}{1+\bar{n}} \right)^{n} \sin^{2}\left(\frac{1}{2}\Omega_{n} t \right), \tag{5.15}$$

where $\bar{n}$ is the mean motional level of the atom. The above equation indicates that the Rabi oscillation can quickly decay due to the distribution in the motional states.

This phenomenon can be simulated by solving the Schrödinger equation with the Hamiltonian

$$\begin{aligned} \mathcal{H} = &- \hbar\delta \, |e\rangle\langle e| \otimes I_{\mathrm{mo}} + \frac{\hbar\Omega_{\mathrm{el}}}{2} \left(|g\rangle\langle e| \, e^{-i\eta(\hat{a}+\hat{a}^{\dagger}}\right) + |e\rangle\langle g| \, e^{i\eta(\hat{a}+\hat{a}^{\dagger})}) \\ &+ I_{\mathrm{el}} \otimes \hbar\omega \sum_{n=0}^{N'} \left(n + \frac{1}{2} \right) |n\rangle\langle n| \, , \end{aligned} \tag{5.16}$$

where δ is the detuning of the laser beam, ω is the trap frequency, N' is the truncation number of the motional states, and I_{el} and I_{mo} are the identity operators for the electronic and motional states, respectively. The operator $\hat{a}$ and $\hat{a}^{\dagger}$ are the annihilation and creation operator for the motional states, respectively. The simulation results are shown in Fig. 5.5. This simulation support that it is necessary to cool the atoms to motional ground state to achieve high-fidelity clock transitions.

We now focus on the experimental implementation of Raman Sideband Cooling (RSC). The RSC process involves driving the red-sideband $|0\rangle \rightarrow |1\rangle$ transition, followed by optical pumping (OP) to return the atoms to the $|0\rangle$ state [1, 11, 12] (Fig. 5.6a). The beam configuration for this process is shown in Fig. 5.4a. The RSC1 and RSC2 beams cool the atoms along the $x - y$ direction (axis 1), while the RSC1 and RSC3 beams cool along the $x + y$ direction (axis 2).

The pulse sequence used for RSC is depicted in Fig. 5.6b. It consists of 16 RSC beam pulses, each with a duration of 6 ms, separated by 2 ms intervals. During this

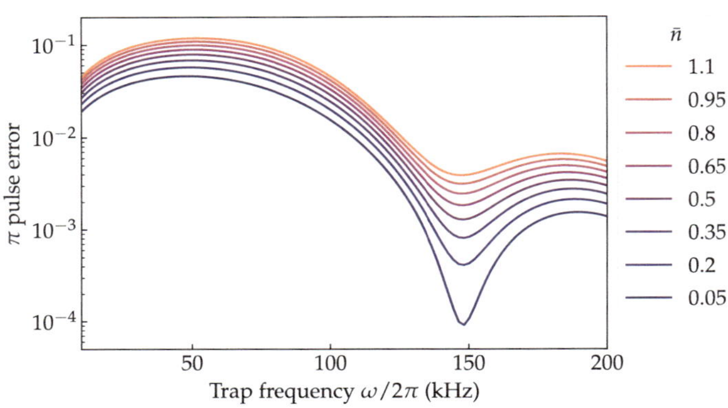

Fig. 5.5 Simulation of clock excitation error. We set the Rabi frequency Ω_{el} to $2\pi \times 50$ kHz in this simulation. The results show that a lower mean motional level significantly enhances the fidelity of clock excitation. The simulation is performed using the QuTiP package [9, 10]

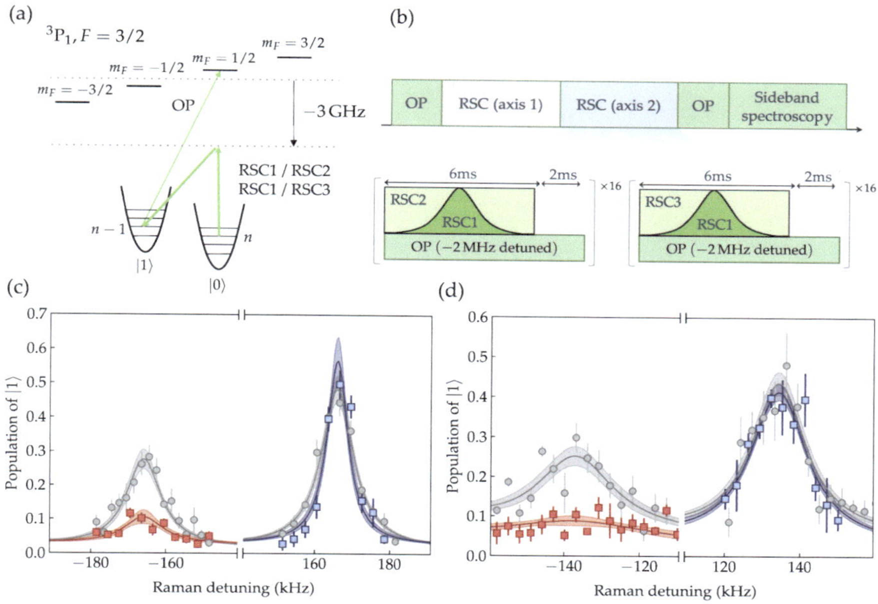

Fig. 5.6 Raman sideband cooling. **a** Relevant energy levels for RSC. **b** Pulse sequence for RSC. **c** and **d** are the Raman sideband spectra along axis 1 and axis 2, respectively. The gray circles represent the data taken without RSC, while the blue and red squares corresponds to that with RSC. The solid lines are the fits to the data using a Lorentzian function, and the shaded areas are 1σ confidence intervals of the fit. The mean motional quantum number is reduced from $\bar{n} = 1.6(4)$ to $\bar{n} = 0.17(4)$ along axis 1 and from $\bar{n} = 1.7(6)$ to $\bar{n} = 0.18(4)$ along axis 2. The error bars represent the standard error of the mean

sequence, the OP beam is simultaneously applied, detuned by −2 MHz from the optimal frequency for OP. After completing the cooling process along axis 1 using the RSC1 and RSC2 beam pair, the RSC sequence is repeated to cool the atoms along axis 2, utilizing the RSC1 and RSC3 beam pair.

To improve the efficiency of RSC, we utilize the HS1 pulse [13, 14], which is defined as

$$\begin{cases} \Omega(t) = \Omega_0 \operatorname{sech}\left[\beta\left(\dfrac{2t}{T} - 1\right)\right], \\ \Delta(t) = \dfrac{\Delta_0}{2} \tanh\left[\beta\left(\dfrac{2t}{T} - 1\right)\right], \end{cases} \tag{5.17}$$

where T is the pulse width, Ω_0 is the peak Rabi frequency, Δ_0 is the frequency sweep range, and β is the pulse shape parameter. This pulse is robust to the inhomogeneity of the trap frequency and enables efficient sideband cooling [15]. In this experiment, we set the pulse shape parameter $\beta = 5$, and the frequency sweep range $\Delta_0 = 2\pi \times 20$ kHz.

Figure 5.6c and d present the results of RSC along axes 1 and 2, respectively. The Raman sideband spectra are taken with a 160 μs pulse of each RSC beam pair, corresponding to the π-pulse for the blue sideband. The mean motional level of the atoms can be estimated from the relation as

$$\frac{A_R}{A_B} = \frac{\bar{n}}{1+\bar{n}} \quad \Longleftrightarrow \quad \bar{n} = \frac{A_R}{A_B - A_R}, \tag{5.18}$$

where A_R and A_B are the amplitudes of the red and blue sidebands, respectively. The mean motional quantum number after the RSC is estimated to be $\bar{n} = 0.17(4)$ along axis 1 and $\bar{n} = 0.18(4)$ along axis 2. These results correspond to a motional ground-state probability of $P_0 = (1+\bar{n})^{-1} = 83\%$. To achieve more efficient cooling, a lower initial atom temperature is required, which can be realized using a magic tweezer array designed for the ${}^1\mathrm{S}_0 \to {}^3\mathrm{P}_1$ transition [4, 5].

5.1.4 Fast Gate Operations on the Nuclear Spin Qubits

The gate operation on the nuclear spin qubits can be significantly accelerated by employing single-beam Raman transitions [1]. In the following, we first provide a theoretical overview of the single-beam Raman transitions. We then present the experimental results of the fast nuclear spin control using single-beam Raman transitions.

The relevant energy levels for the single-beam Raman transitions, which are labeled as $|0\rangle$, $|1\rangle$, $|2\rangle = \left|{}^3\mathrm{P}_1, F = 3/2, m_F = -1/2\right\rangle$, and $|3\rangle = \left|{}^3\mathrm{P}_1, F = 3/2, m_F = 1/2\right\rangle$, are shown in Fig. 5.7a. Note that this labeling implicitly assumes that the detuning of the laser beam is not so far from the ${}^3\mathrm{P}_1$, $F = 3/2$ manifolds, and thus the ${}^3\mathrm{P}_1$, $F = 1/2$ manifolds are not involved in the Raman transitions. The Hamiltonian for the single-beam Raman transitions is given by

$$\mathcal{H}/\hbar = \begin{pmatrix} \Delta_N & 0 & \Omega_{02}/2 & \Omega_{03}/2 \\ 0 & 0 & \Omega_{12}/2 & \Omega_{13}/2 \\ \Omega_{02}^*/2 & \Omega_{12}^*/2 & \Delta_X & 0 \\ \Omega_{03}^*/2 & \Omega_{13}^*/2 & 0 & \Delta_X + \Delta_e \end{pmatrix}, \tag{5.19}$$

where Δ_N is the splitting between the $|0\rangle$ and $|1\rangle$ states, Δ_e is the splitting between the $|2\rangle$ and $|3\rangle$ states, Δ_X is the detuning of the laser beam from the $|1\rangle \leftrightarrow |2\rangle$ transition, and Ω_{ij} is the Rabi frequency between the $|i\rangle$ and $|j\rangle$ states. If we take a quantization axis along the z-axis and the laser beam direction along the x-axis, the electric field of the laser beam can be written as $\boldsymbol{E} = E_y \cos(\omega t + \phi_{H/V})\hat{\boldsymbol{y}} + E_z \cos(\omega t)\hat{\boldsymbol{z}}$ with the relative phase between the horizontal and vertical polarization $\phi_{H/V}$. The Hamiltonian is then rewritten as

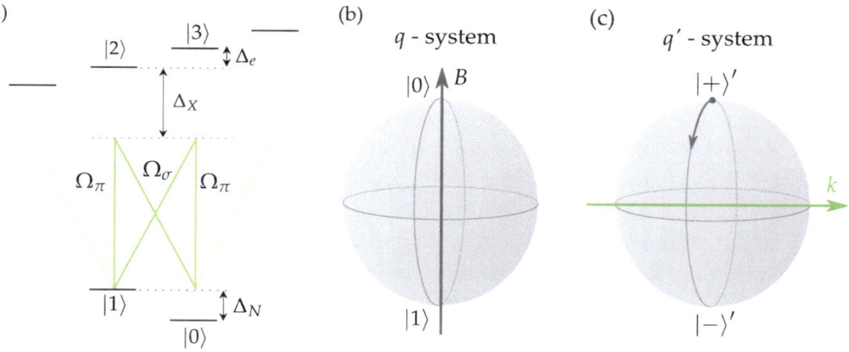

Fig. 5.7 Single-beam Raman transition. **a** Relevant energy levels for the single-beam Raman transitions. **b** Bloch sphere representation of the nuclear spin states. The quantization axis is determined by the magnetic field. **c** Bloch sphere representation of the nuclear spin states in the q'-system. The strong gate beam defines the quantization axis. Consequently, the $|0\rangle$ state in the q-system is represented as the $|+\rangle'$ state in the q'-system and rotates around the x-axis with an angular frequency of Ω_R

$$\mathcal{H}/\hbar = \begin{pmatrix} \Delta_N & 0 & ie^{i\phi_{H/V}}\Omega_\sigma/2 & \Omega_\pi/2 \\ 0 & 0 & \Omega_\pi/2 & ie^{i\phi_{H/V}}\Omega_\sigma/2 \\ -ie^{-i\phi_{H/V}}\Omega_\sigma/2 & \Omega_\pi^*/2 & \Delta_X & 0 \\ \Omega_\pi^*/2 & -ie^{-i\phi_{H/V}}\Omega_\sigma^*/2 & 0 & \Delta_X + \Delta_e) \end{pmatrix}, \quad (5.20)$$

with π-transition strength of Ω_π and σ-transition strength of Ω_σ. By adiabatically eliminating the excited states and assuming $\Delta_X \gg \Delta_e$, we obtain the effective Hamiltonian for the nuclear spin states as

$$\mathcal{H}_{\text{eff}}/\hbar = -\frac{1}{4\Delta_X}\begin{pmatrix} -4\Delta_N\Delta_X + \Omega_\pi^2 + \Omega_\sigma^2 & 2\sin(\phi_{H/V})\Omega_\pi\Omega_\sigma \\ 2\sin(\phi_{H/V})\Omega_\pi\Omega_\sigma & \Omega_\pi^2 + \Omega_\sigma^2) \end{pmatrix}. \quad (5.21)$$

Therefore, the Rabi frequency of the single-beam Raman transition is given by

$$\Omega_R = \left|\sin(\phi_{H/V})\right| \Omega_\pi\Omega_\sigma/\Delta_X. \quad (5.22)$$

This quantity is maximized when $\phi_{H/V} = \pm\pi/2$, that is, when the laser beam is circulary polarized.

Here, we provide an intuitive interpretation of the single-beam Raman transitions. Consider a nuclear spin polarized along the z-axis. In this scenario, the quantization axis (hereafter referred to as the q-system) is defined by the magnetic field. The Bloch sphere representation of the nuclear spin states is shown in Fig. 5.7b. When a circularly polarized single-beam Raman laser is applied, the nuclear spin states experience state-dependent light shifts as given by Eq. (5.11), resulting in a nuclear spin splitting larger than the Zeeman splitting.

In this situation, the quantization axis is effectively determined by the direction of the laser beam (referred to as the q'-system). As a result, the $|0\rangle$ state in the q-system is represented as the $|+\rangle'$ state in the q'-system (Fig. 5.7c). The state then rotates around the laser beam direction (x-axis) with an angular frequency of $\Omega_R = |U_{1/2} - U_{-1/2}|/\hbar$. This rotation can be interpreted as the Raman transition between the $|0\rangle$ and $|1\rangle$ states, corresponding to the implementation of the X gate.

We now present experiment results of the fast nuclear spin control of X and Z gates. The formar is performed by employing single beam Raman transition described above, while the latter is by applying a circularly polarized laser beam along the quantization axis. The beam configuration is shown in Fig. 5.8a. The beams are focused onto the atom arrays with a beam radii of 45 μm in the vertical direction and 600 μm in the horizontal direction. The quantization axis is set along the Z gate beam direction (y-axis) by applying a magnetic field of 2.5 G. The gate beams are orthogonal to each other.

Figure 5.8b present the experimental result of the X gate. The beam power is 105 mW and have a detuning of –3 GHz from the $^3P_1$, $F = 3/2$ manifolds. We successfully observe the Rabi oscillation between the $|0\rangle$ and $|1\rangle$ states with a Rabi frequency of $2\pi \times 1.27$ MHz. The fidelity of π-pulse $X(\pi)$ is found to be approximately 0.995 by comparing the survival probabilities with and without blasting out the atoms in $|0\rangle$ after applying a π-pulse. This fidelity might be limited by the inhomogeneity of the beam intensity over the atom array and the shot-to-shot fluctuation of the pulse area.

Figure 5.8c shows the experimental result of the Z gate. We first apply $X(\pi/2)$ gate to prepare the superposition state of $|0\rangle$ and $|1\rangle$. We then irradiate the Z gate beam, followed by the additional $X(\pi/2)$ pulse to measure the population of the $|0\rangle$ state. The oscillation with a frequency of $2\pi \times 117$kHz does not exhibit any damping

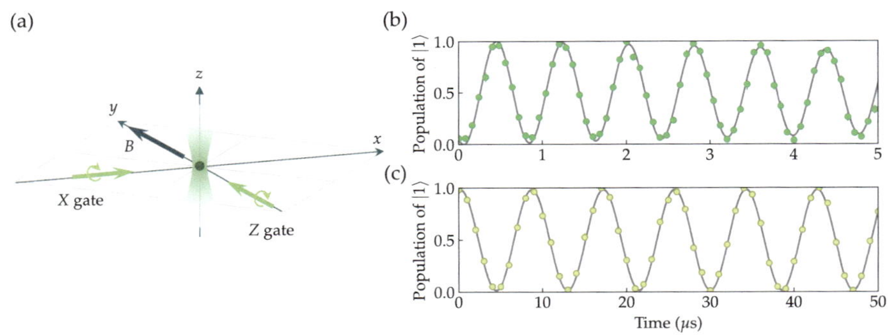

Fig. 5.8 Gate operations on the nuclear spin qubits. **a** Schematic of the beam configuration for the X and Z gates. **b** Rabi oscillations between the $|0\rangle$ and $|1\rangle$ states driven by the X gate beam, with a Rabi frequency of $2\pi \times 1.27$ MHz. The solid line is the fit to the data using a damped sinusoidal function. **c** Phase rotation by the Z gate beam with a oscillation frequency of $2\pi \times 117$ kHz. The solid line is the fit to the data using a damped sinusoidal function. The error bars represent the standard error of the mean

behavior within the measured range, indicating that the Z gate is implemented with high fidelity. The fidelity of the Z gate is estimated to be 0.999 by comparing the survival probabilities with and without the Z gate beam irradiation and compensating the infidelity of the $X(\pi/2)$ gates. More regorous evaluation of the gate fidelity will be performed with randomized benchmarking technique [16].

5.1.4.1 Optimal Detuning for High-Fidelity Gate Operations

Since photon scattering during gate operation can lead to infidelity, it is crucial to carefully select the detuning of the single-qubit gate beam to optimize the fidelity. One straightforward way to estimate the gate operation fidelity is to evaluate the ratio between the Rabi frequency and the scattering rate at a given detuning. The Rabi frequency of the single-beam Raman transition corresponds to the difference in light shifts between the $|0\rangle$ and $|1\rangle$ states, which is derived from Eq. (5.11) as

$$\Omega_R = |U_{1/2} - U_{-1/2}|/\hbar \propto \frac{1}{\Delta^2} \text{ (for } \Delta \gg \Delta_{\text{HF}}), \tag{5.23}$$

where Δ_{HF} represents the hyperfine splitting of the excited state.

On the other hand, the scattering rate is given by Eq. (5.12) as

$$\gamma_{\text{sc}} \propto \frac{1}{\Delta^2}. \tag{5.24}$$

At first glance, this suggests that the ratio between the Rabi frequency and the scattering rate is independent of detuning. However, this conclusion is incorrect because the scattering rate in Eq. (5.12) is not entirely accurate, as previously noted. The scattering rate must account for both Raman and Rayleigh scattering. The former is described by the Kramers-Heisenberg formula [17] as

$$\gamma_{ij} = g^2\Gamma \sum_q \left(\sum_{F',m_{F'}} A^{i,j}_{F',m_{F'},q} \right)^2, \tag{5.25}$$

while the latter is expressed as

$$\gamma_{\text{el}} = g^2\Gamma \sum_q \left[\sum_{F',m_{F'}} \left(A^{+1/2,-1/2}_{F',m_{F'},q} - A^{-1/2,+1/2}_{F',m_{F'},q} \right) \right]^2. \tag{5.26}$$

Here, $g = dE_0/(2\hbar)$ and $d = (3\pi\epsilon_0\hbar c^3\Gamma/\omega_0^3)^{1/2}$, where ω_0 is the transition frequency, and the electric field of the light is $E_0 = \sum_q b_q \hat{\epsilon_q}$. The scattering amplitude for transitions between nuclear spin states $|i\rangle$ and $|j\rangle$ is given by

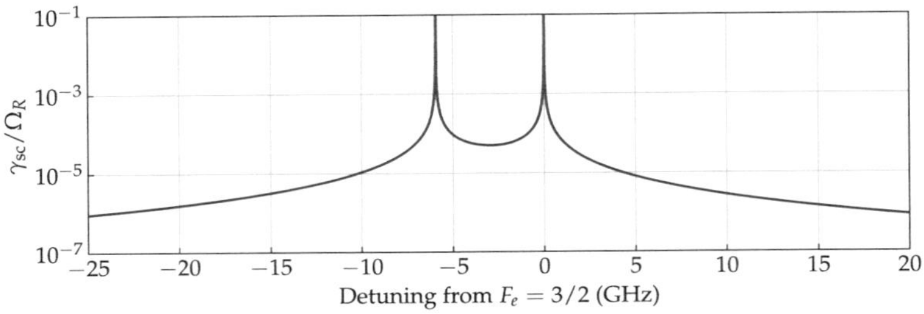

Fig. 5.9 Detuning dependency of the ratio of the scattering rate to the Rabi frequency

$$A^{i,j}_{F',m_{F'},q} = \frac{b_q \left\langle j \middle| \hat{d}_q \cdot \hat{\epsilon}^*_{q+(i-j)} \middle| F', m_{F'} \right\rangle \left\langle F', m_{F'} \middle| \hat{d}_q \cdot \hat{\epsilon}_q \middle| i \right\rangle}{\Delta_{F',m_{F'}} d^2}, \tag{5.27}$$

where $\Delta_{F',m_{F'}}$ is the detuning from the intermediate state $|F', m_{F'}\rangle$.

For the equal superposition state $|+\rangle = (|0\rangle + |1\rangle)/\sqrt{2}$, the total scattering rate becomes $\gamma_{sc} = (\gamma_{+1/2,-1/2} + \gamma_{-1/2,+1/2} + \gamma_{el})/2$. The ratio of the scattering rate to the Rabi frequency is then expressed as [1]

$$\frac{\gamma_{sc}}{\Omega_R} \simeq \frac{\Gamma}{\sqrt{6}} \left| \frac{\Delta_{HF}}{\Delta(\Delta - \Delta_{HF})} \right|, \tag{5.28}$$

where Δ is the detuning from the $F_e = 3/2$ manifolds of the $^3P_1$ state. Figure 5.9 shows the ratio of the scattering rate to the Rabi frequency as a function of the detuning. When the detuning lies between the hyperfine splitting, the ratio is locally minimized. Considering the Rabi frequency at a given detuning (Fig. 5.3), this detuning range is preferable. Further suppression of scattering can be achieved by increasing the detuning, albeit at the cost of a reduced Rabi frequency. In Yb system, a single qubit gate fidelity of 0.99968(3) is achieved with a detuning of –5.8 GHz from the $^3P_1$, $F' = 1/2$ manifolds.

5.1.5 Coherence Time Evaluation of the Nuclear Spin Qubits

The coherence time of the nuclear spin qubits is evaluated by using the Ramsey sequence. The sequence consists of two $X(\pi/2)$ pulses separated by a variable free evolution time T. The Ramsey signal is obtained by measuring the population of the $|1\rangle$ state after the second $X(\pi/2)$ pulse. The coherence time is then extracted by fitting the Ramsey signal to a damped sinusoidal function. Figure 5.10 shows the experimental results of the Ramsey sequence. In this measurement, we set the magnetic field to 2.5 G along y-axis and the trap potential to 35 μK. The coherence

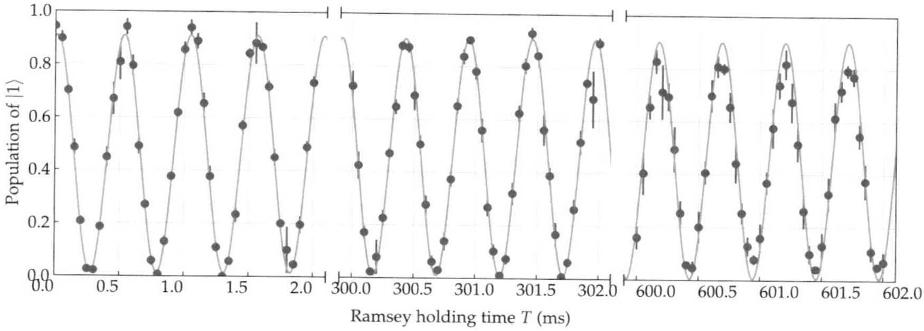

Fig. 5.10 Coherence time evaluation of the nuclear spin qubit. The Ramsey signal is obtained by measuring the population of the $|1\rangle$ state after the second $X(\pi/2)$ pulse. The solid line is the fit to the data using a damped sinusoidal function. The coherence time is found to be $T_2^* = 3.8(4)$ s. The error bars represent the standard error of the mean

time is estimated to be $T_2^* = 3.8(4)$ s. The coherence time is limited by the magnetic field noise. The coherence time can be improved by using dynamical decoupling technique [18].

5.1.6 Coherent ^{3}P$_2$ Excitation of ^{171}Yb

As a first step towards Rydberg excitation, we demonstrate the coherent excitation of the ^{3}P$_2$ state in ^{171}Yb. We begin by cooling the motional states of the atoms using the RSC technique described in the previous section. Subsequently, we adiabatically reduce the magnetic field from 48 to 1 G and lower the trapping depth to 35 μK to minimize the differential light shift between the ground ^{1}S$_0$ and the metastable ^{3}P$_2$ states. The excitation is then performed using a 507 nm beam, with beam radii of 35 μm in the vertical direction and 500 μm in the horizontal direction, propagating along the x-axis. The polarization of the excitation beam is set to horizontal, enabling the HFM-E1 $|0\rangle \leftrightarrow |^3\mathrm{P}_2, F = 3/2, m_F = 3/2\rangle$ transition under the z-axis magnetic field.

Figure 5.11a shows the spectroscopy result for the $|^3\mathrm{P}_2, F = 3/2, m_F = 3/2\rangle$ state. The beam power is 1.5 mW, and the spectroscopy pulse duration is set to 400 μs. The observed linewidth is 2.6(4) kHz, primarily limited by magnetic field fluctuations during spectroscopy. Increasing the beam power to 35 mW, we perform coherent excitation. Figure 5.11b displays the Rabi oscillation between the $|0\rangle$ and $|^3\mathrm{P}_2, F = 3/2, m_F = 3/2\rangle$ states, with a measured Rabi frequency of $2\pi \times 9.1(2)$ kHz.

The low excitation efficiency to the ^{3}P$_2$ state is attributed to the differential light shift between the ^{1}S$_0$ and ^{3}P$_2$ states. This efficiency can be enhanced by employing an adiabatic rapid passage technique or a magic tweezer array optimized for the ^{1}S$_0 \leftrightarrow {}^3$P$_2$ transition.

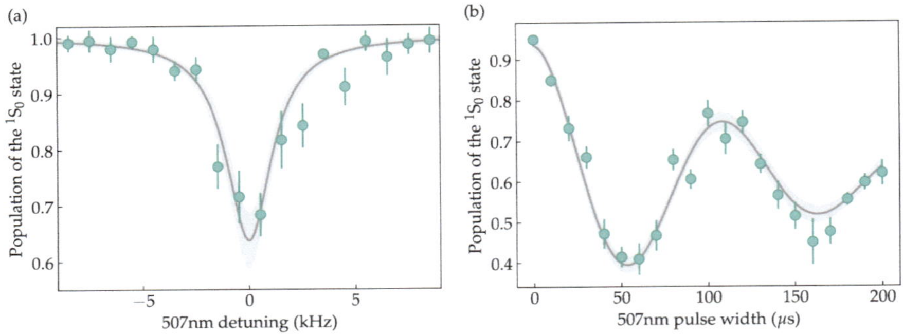

Fig. 5.11 Spectroscopy and coherent excitation of the $^3P_2$ state of ^{171}Yb. **a** Spectrum of the $|^3P_2, F = 3/2, m_F = 3/2\rangle$ state with a linewidth of 2.6(4) kHz. The solid line is the fit using a Lorentzian function. The shaded area represents 1σ confidence interval of the fit. **b** Rabi oscillation between the $|0\rangle$ and $|^3P_2, F = 3/2, m_F = 3/2\rangle$ states with a Rabi frequency of $2\pi \times 9.1(2)$ kHz. The solid line is the fit using a damped sine function. The error bars represent the standard error of the mean

5.2 Optical Clock Qubits in ^{174}Yb

5.2.1 Basic Properties

The optical clock qubits encoded in the ground state and a metastable state of bosonic alkaline-earth(-like) atoms are promising candidates for ancilla qubits due to several advantageous properties. First, the qubit states with $m_J = 0$ are highly insensitive to magnetic field fluctuations, which is crucial for preserving qubit coherence. Indeed, second-order coherence times have been reported in previous studies [19, 20]. Second, state-selective qubit readout without necessity of state-selective atom loss or requiring specialized techniques such as shelving is naturally feasible, as the metastable state serves as a dark state for the probe transition. This unique feature makes these qubits particularly appealing as ancilla qubits.

In this study, we utilize an optical clock qubit encoded as $|0\rangle = |^1S_0\rangle$ and $|1\rangle = |^3P_2, m_J = 0\rangle$. This choice is motivated by the fact that our tweezer wavelength of 532 nm is near the magic wavelength for these qubit states, minimizing the differential light shift. Furthermore, the transition strength of the $^1S_0 \leftrightarrow {}^3P_2$ transition is significantly larger than that of the $^1S_0 \leftrightarrow {}^3P_1$ transition, enabling faster gate operations. This is because the former is an M2 transition, while the latter relies on magnetic-field-induced state mixing [21]. Additionally, the $^3P_2$ state is less sensitive to magnetic field fluctuations compared to the $^3P_1$ state [22], which is advantageous for maintaining qubit coherence.

5.2.2 Narrow-Line Spectroscopy of the $^3P_2$ State

We first perform spectroscopy of the $^3P_2$ state to evaluate the transition linewidth in our setup. The excitation beam at 507 nm is applied from the horizontal direction with vertical polarization. The magnetic field is oriented at an angle of 45° relative to the excitation laser beam, and the tweezer polarization is aligned to the magnetic field. This configuration maximizes the transition strength of the $^1S_0 \leftrightarrow {}^3P_2, m_J = 0$ transition (see Fig. 2.2). The beam power is set to 3 mW, and the spectroscopy pulse duration is 10 ms.

We repeat the spectroscopy at various trap depths ranging from 35.4 to 212.4 μK to investigate the ratio of polarizabilities between the qubit states. The results are shown in Fig. 5.12. The slope of the resonance shift as a function of trap depth is expressed as $A = 1 - \alpha_e/\alpha_g$, where α_e and α_g are the polarizabilities of the $^3P_2$ and $^1S_0$ states, respectively. From the linear fit, we obtain $A = -0.142(1)$ kHz/μK, corresponding to $\alpha_e/\alpha_g = 1.0068(1)$. This result is consistent with the theoretical prediction of $\alpha_e/\alpha_g = 1.016$ (Fig. 4.3). Note that we use the theoretical ground-state polarizability of $\alpha_g = 37.9\,\mathrm{Hz/(W/cm^2)}$ in this calculation. Without any theoretical input, we estimate the differential polarizability as $\Delta\alpha = \alpha_e - \alpha_g = -6.44(5) \times 10^{-2}\,\mathrm{Hz/(W/cm^2)}$ using the tweezer beam radii and the beam power.

The spectrum of the $^3P_2$ state in the 35.4 μK trap is shown in Fig. 5.13. The observed linewidth is 1.3(2) kHz, which is narrower than the linewidth of the $^3P_2$ state in ^{171}Yb (2.6(4) kHz), owing to its insensitivity to the magnetic field fluctuations and near-magic condition. However, the linewidth is still too broad considering the differential polarizability of $\Delta\alpha = -6.44(5) \times 10^{-2}\,\mathrm{Hz/(W/cm^2)}$, described above. This indicates that there can be additional sources of the linewidth broadening. We leave the detailed investigation of the broadening to future work.

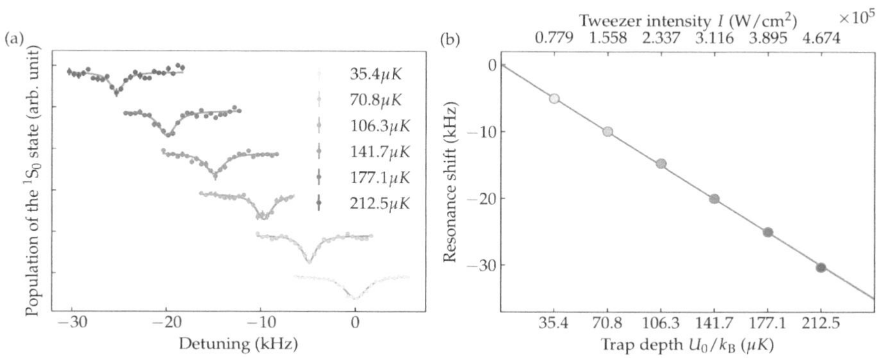

Fig. 5.12 Measurement of the differential light shift between the $^1S_0$ and the $^3P_2$ states of ^{174}Yb. **a** The spectra taken at various trap depths. The solid lines are the fits to the data using a Lorentzian function. **b** The resonance shift as a function of the trap depth is shown. The solid line is the fit to the data using a linear function. The differential polarizability is estimated to be $\Delta\alpha = -6.44(5) \times 10^{-2}\,\mathrm{Hz/(W/cm^2)}$. The error bars represent the standard error of the mean

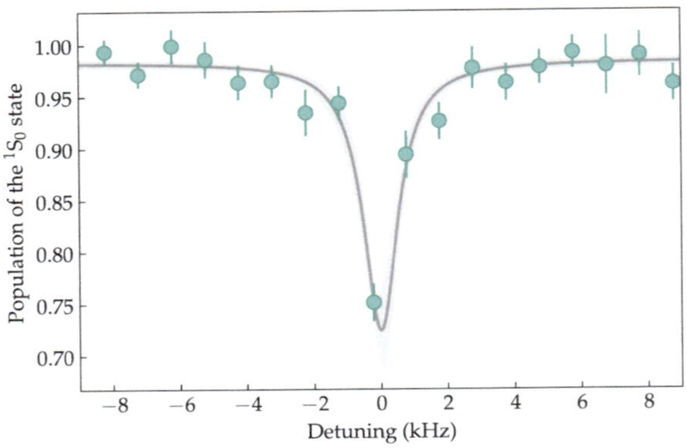

Fig. 5.13 Narrow-line spectroscopy of the $^3P_2$ state of ^{174}Yb. The spectrum is taken at a trap depth of 35.4 μK. The solid line is the fit to the data using a Lorentzian function, yielding the linewidth of 1.3(2) kHz. The shaded area represents 1σ confidence interval of the fit. The error bars represent the standard error of the mean

5.2.3 *Coherent Control and Coherence Time Measurement of the Optical Clock Qubit*

We next perform coherent control of the optical clock qubit. For this, we narrow the excitation beam radii to 45 μm in vertical direction and 105 μm in horizontal direction, and increase the beam power to 210 mW to obtain enough intensity. Figure 5.14a shows the Rabi oscillation between the $|0\rangle$ and $|1\rangle$ states in a 35.4 μk tweezer array. The Rabi frequency is measured to be $2\pi \times 9.5(2)$ kHz. It should be noted that the

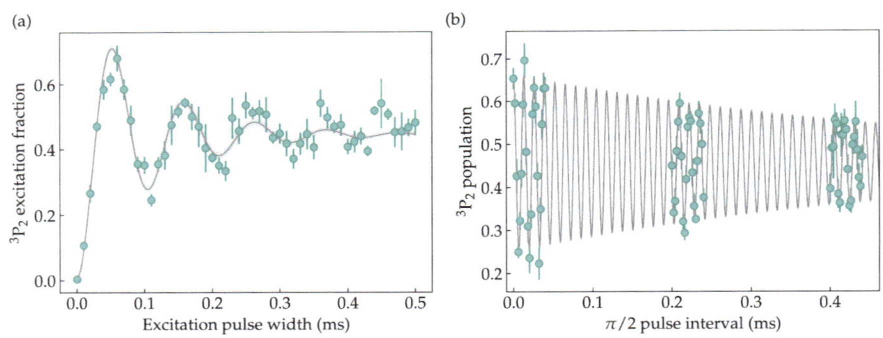

Fig. 5.14 Coherent manipulation of the optical clock qubits. **a** Rabi oscillation between the ground and the $^3P_2$ states. The gray curve is a fitting function with a damping sine function, yielding a Rabi frequency of $2\pi \times 9.5(2)$ kHz. The shaded area is 1σ-confidence intervals of the fitting. The error bars represent the standard error of the mean. **b** The coherence of the optical clock qubit measured by the Ramsey sequence. From the fitting with a damping sine function, the coherence time is estimated to be 0.55(7) ms. The error bars represent the standard error of the mean

Rabi oscillation is primarily damped by the thermal distribution of the motional states in the tweezer potential [8]. In this experiment, the mean vibrational quantum number $\bar{n}$ is approximately $\bar{n} = 1$, resulting in variations in the Rabi frequency due to motional coupling. This can be addressed through sideband cooling to the motional ground state via the clock transition [23].

Furthermore, we evaluate the coherence of the optical clock qubits using a Ramsey sequence. Figure 5.14b shows the Ramsey signal obtained by scanning the holding time between the two $\pi/2$ pulses on the $^1S_0 \leftrightarrow {}^3P_2$ transition. Fitting the data with a damped sine function yields a coherence time of 0.55(7) ms. The oscillation frequency of 76.55(4) kHz corresponds to the probe light shift induced by the 507 nm beam. The coherence time might be currently shortened due to the spatial distribution of the differential light shift between the ground and $^3P_2$ states in each tweezer potential [24], as well as the aforementioned thermal distribution, which induces slight shot-to-shot variations in the Ramsey frequency. Meanwhile, we anticipate achieving a coherence time on the order of several milliseconds for the $^3P_2$ optical clock qubits in magic-wavelength traps, as recently demonstrated in a strontium system [25].

References

1. A. Jenkins, J.W. Lis, A. Senoo, W.F. McGrew, A.M. Kaufman, Phys. Rev. X **12**(2), 021027 (2022). https://doi.org/10.1103/PhysRevX.12.021027. https://link.aps.org/doi/10.1103/PhysRevX.12.021027
2. S. Ma, A.P. Burgers, G. Liu, J. Wilson, B. Zhang, J.D. Thompson, Phys. Rev. X **12**(2), 021028 (2022). https://doi.org/10.1103/PhysRevX.12.021028. https://link.aps.org/doi/10.1103/PhysRevX.12.021028
3. S. Ma, G. Liu, P. Peng, B. Zhang, S. Jandura, J. Claes, A.P. Burgers, G. Pupillo, S. Puri, J.D. Thompson, Nature (London) **622**(7982), 279 (2023). https://doi.org/10.1038/s41586-023-06438-1. http://dx.doi.org/10.1038/s41586-023-06438-1
4. J.W. Lis, A. Senoo, W.F. McGrew, F. Rönchen, A. Jenkins, A.M. Kaufman, Phys. Rev. X **13**(4), 041035 (2023). https://doi.org/10.1103/PhysRevX.13.041035. https://link.aps.org/doi/10.1103/PhysRevX.13.041035
5. M.A. Norcia, W.B. Cairncross, K. Barnes, P. Battaglino, A. Brown, M.O. Brown, K. Cassella, C.A. Chen, R. Coxe, D. Crow, J. Epstein, C. Griger, A.M.W. Jones, H. Kim, J.M. Kindem, J. King, S.S. Kondov, K. Kotru, J. Lauigan, M. Li, M. Lu, E. Megidish, J. Marjanovic, M. McDonald, T. Mittiga, J.A. Muniz, S. Narayanaswami, C. Nishiguchi, R. Notermans, T. Paule, K.A. Pawlak, L.S. Peng, A. Ryou, A. Smull, D. Stack, M. Stone, A. Sucich, M. Urbanek, R.J.M. van de Veerdonk, Z. Vendeiro, T. Wilkason, T.Y. Wu, X. Xie, X. Zhang, B.J. Bloom, Phys. Rev. X **13**(4), 041034 (2023). https://doi.org/10.1103/PhysRevX.13.041034. https://link.aps.org/doi/10.1103/PhysRevX.13.041034
6. J.A. Muniz, M. Stone, D.T. Stack, M. Jaffe, J.M. Kindem, L. Wadleigh, E. Zalys-Geller, X. Zhang, C.A. Chen, M.A. Norcia, J. Epstein, E. Halperin, F. Hummel, T. Wilkason, M. Li, K. Barnes, P. Battaglino, T.C. Bohdanowicz, G. Booth, A. Brown, M.O. Brown, W.B. Cairncross, K. Cassella, R. Coxe, D. Crow, M. Feldkamp, C. Griger, A. Heinz, A.M.W. Jones, H. Kim, J. King, K. Kotru, J. Lauigan, J. Marjanovic, E. Megidish, M. Meredith, M. McDonald, R. Morshead, S. Narayanaswami, C. Nishiguchi, T. Paule, K.A. Pawlak, K.L. Pudenz, D.R. Pérez, A. Ryou, J. Simon, A. Smull, M. Urbanek, R.J.M. van de Veerdonk, Z. Vendeiro, T.Y. Wu, X. Xie, B.J. Bloom (2024). arXiv:2411.11708

7. D. Bluvstein, H. Levine, G. Semeghini, T.T. Wang, S. Ebadi, M. Kalinowski, A. Keesling, N. Maskara, H. Pichler, M. Greiner, V. Vuletić, M.D. Lukin, Nature (London) **604**(7906), 451 (2022). https://doi.org/10.1038/s41586-022-04592-6. https://www.nature.com/articles/s41586-022-04592-6
8. M. Takamoto, H. Katori, J. Phys. Soc. Jpn. **78**(1), 013301 (2009). https://doi.org/10.1143/JPSJ.78.013301
9. J.R. Johansson, P.D. Nation, F. Nori, Comput. Phys. Commun. **183**(8), 1760 (2012). http://dx.doi.org/10.1016/j.cpc.2012.02.021
10. J.R. Johansson, P.D. Nation, F. Nori, Comput. Phys. Commun. **184**(4), 1234 (2013). http://dx.doi.org/10.1016/j.cpc.2012.11.019
11. A.M. Kaufman, B.J. Lester, C.A. Regal, Phys. Rev. X **2**(4), 041014 (2012).https://doi.org/10.1103/PhysRevX.2.041014. https://link.aps.org/doi/10.1103/PhysRevX.2.041014
12. J.D. Thompson, T.G. Tiecke, A.S. Zibrov, V. Vuletić, M.D. Lukin, Phys. Rev. Lett. **110**(13), 133001 (2013). http://dx.doi.org/10.1103/PhysRevLett.110.133001
13. J. Baum, R. Tycko, A. Pines, Phys. Rev. A Gen. Phys. **32**(6), 3435 (1985). https://doi.org/10.1103/physreva.32.3435. https://link.aps.org/doi/10.1103/PhysRevA.32.3435
14. M. Garwood, L. DelaBarre, J. Magn. Reson. **153**(2), 155 (2001). http://dx.doi.org/10.1006/jmre.2001.2340
15. Y. Chew, T. Tomita, T.P. Mahesh, S. Sugawa, S. de Léséleuc, K. Ohmori, Nat. Photonics 1–6 (2022). https://doi.org/10.1038/s41566-022-01047-2. https://www.nature.com/articles/s41566-022-01047-2
16. J. Emerson, R. Alicki, K. Życzkowski, J. Opt. B Quantum Semiclassical Opt. **7**(10), S347 (2005). https://doi.org/10.1088/1464-4266/7/10/021. https://iopscience.iop.org/article/10.1088/1464-4266/7/10/021/meta
17. H. Uys, M.J. Biercuk, A.P. Vandevender, C. Ospelkaus, D. Meiser, R. Ozeri, J.J. Bollinger, Phys. Rev. Lett. **105**(20), 200401 (2010). https://doi.org/10.1103/PhysRevLett.105.200401. http://link.aps.org/pdf/10.1103/PhysRevLett.105.200401
18. G.T. Genov, D. Schraft, N.V. Vitanov, T. Halfmann, Phys. Rev. Lett. **118**(13), 133202 (2017). http://dx.doi.org/10.1103/PhysRevLett.118.133202
19. M.A. Norcia, A.W. Young, W.J. Eckner, E. Oelker, J. Ye, A.M. Kaufman, Science **366**(6461), 93 (2019). http://dx.doi.org/10.1126/science.aay0644
20. A.W. Young, W.J. Eckner, W.R. Milner, D. Kedar, M.A. Norcia, E. Oelker, N. Schine, J. Ye, A.M. Kaufman, Nature **588**(7838), 408 (2020). http://dx.doi.org/10.1038/s41586-020-3009-y
21. A.V. Taichenachev, V.I. Yudin, C.W. Oates, C.W. Hoyt, Z.W. Barber, L. Hollberg, Phys. Rev. Lett. **96**(8), 083001 (2006). http://dx.doi.org/10.1103/PhysRevLett.96.083001
22. V.A. Dzuba, V.V. Flambaum, S. Schiller, Phys. Rev. A **98**(2), 022501 (2018). https://doi.org/10.1103/PhysRevA.98.022501. https://link.aps.org/doi/10.1103/PhysRevA.98.022501
23. X. Zhang, K. Beloy, Y.S. Hassan, W.F. McGrew, C.C. Chen, J.L. Siegel, T. Grogan, A.D. Ludlow, Phys. Rev. Lett. **129**(11), 113202 (2022). https://doi.org/10.1103/PhysRevLett.129.113202. https://link.aps.org/doi/10.1103/PhysRevLett.129.113202
24. G. Unnikrishnan, P. Ilzhöfer, A. Scholz, C. Hölzl, A. Götzelmann, R.K. Gupta, J. Zhao, J. Krauter, S. Weber, N. Makki, H.P. Büchler, T. Pfau, F. Meinert, Phys. Rev. Lett. **132**(15), 150606 (2024). http://dx.doi.org/10.1103/PhysRevLett.132.150606
25. V. Klüsener, S. Pucher, D. Yankelev, J. Trautmann, F. Spriestersbach, D. Filin, S.G. Porsev, M.S. Safronova, I. Bloch, S. Blatt, Phys. Rev. Lett. **132**(25), 253201 (2024). https://doi.org/10.1103/physrevlett.132.253201. https://journals.aps.org/prl/abstract/10.1103/PhysRevLett.132.253201

Chapter 6
Hybrid Atom Tweezer Array of Nuclear Spin and Optical Clock Qubits

This chapter reports on the experimental demonstration of a hybrid atom tweezer array of nuclear spin and optical clock qubits based on the published paper [1]. The results show that this system enables low crosstalk measurement of the optical clock qubits while maintaining the nuclear spin qubits, which is a key step towards mid-circuit measurements for quantum error correction.

6.1 Challenges with Mid-Circuit Measurements in Neutral Atom Arrays

Ancilla qubits are pivotal in quantum error correction (QEC), a cornerstone for achieving fault-tolerant quantum computing [2]. The effectiveness of ancilla-based QEC protocols has been recently validated in quantum computing platforms utilizing superconducting qubits [3–5] and trapped ion qubits [6, 7], underscoring their critical role in advancing quantum technologies.

Simultaneously, the development of optical tweezer array systems has seen remarkable progress, enabling the realization of large-scale qubit arrays [8–13] and high-fidelity gate operations [14–19]. These systems offer unique advantages, such as dynamic qubit connectivity facilitated by moving tweezers [20, 21] and error-detection techniques leveraging the rich internal states of atoms [22], paving the way for implementing QEC protocols [23–27].

A key challenge in optical tweezer array systems is the efficient control and readout of ancilla qubits while minimizing qubit loss and crosstalk with neighboring data qubits. This challenge primarily arises from the use of globally applied light for these operations, which can lead to heating and decoherence of the atoms [28].

To address these issues, two main strategies have been explored: spatial and spectral isolation of data qubits from the ancilla qubit control light. Spatial isolation leverages moving tweezers to transport qubits between distinct zones designated for

Y. Nakamura, *Generation of Highly Controllable Ytterbium Atom Array*,
Springer Theses, https://doi.org/10.1007/978-981-95-2836-3_6

specific operations [21, 29]. Spectral isolation employs local addressing beams to induce light shifts, often combined with advanced state manipulation, to make data qubit atoms in a dark state [30–32].

An alternative strategy involves utilizing two distinct qubits within a single optical tweezer array, where one serves as a data qubit and the other as an ancilla qubit. Dual-species [35–37] and dual-isotope [38] alkali atom arrays have demonstrated the feasibility of qubit readout with minimal crosstalk, thanks to their well-separated resonant frequencies.

An ideal system for ancilla-based QEC protocols [39] would feature data qubits with inherently long coherence times and ancilla qubits capable of non-destructive measurements, while minimizing crosstalk and avoiding the need for complex additional techniques. To this end, we present a hybrid system based on a dual-isotope alkaline-earth-like ytterbium (Yb) atom array. In this system, the fermionic isotope ^{171}Yb provides a nuclear spin qubit as a data qubit, offering long coherence times and fast qubit control [40, 41], while the bosonic isotope ^{174}Yb serves as an ancilla qubit, utilizing an optical clock transition for non-destructive, state-selective qubit readout [42] (Fig. 6.1a). The following sections detail the experimental setup and results of this hybrid atom tweezer array of nuclear spin and optical clock qubits.

6.2 Preparation of a Dual-Yb Atom Array

Our experimental sequence begins with the simultaneous trapping of both isotopes in a MOT (Fig. 6.1b). The atoms are then loaded into a 10×10 tweezer array with a depth of $k_B \times 0.9$ mK and a beam radius of approximately 550 nm. The tweezer wavelength is 532 nm, which is near the magic wavelength for both the $^1S_0 - {}^3P_1$ transition [43, 44] and the $^1S_0 - {}^3P_2$ transition [45, 46] of ^{174}Yb. After applying red-detuned LAC beams at 556 nm for each isotope to prepare single atoms in the tweezers, we image them sequentially with a 20 ms exposure time. This is achieved by irradiating a single resonant 399 nm probe beam while simultaneously cooling with MOT beams (Fig. 4.6). The probe and cooling beam intensities are $1 \times 10^{-3}\ I_{s,399}$ and $10\ I_{s,556}$, respectively, where $I_{s,399} = 59\,\mathrm{mW/cm^2}$ and $I_{s,556} = 0.14\mathrm{mW/cm^2}$ are the saturation intensities of the $^1S_0 - {}^1P_1$ and $^1S_0 - {}^3P_1$ transitions, respectively. The measured atom temperatures are 38(2) μK for ^{171}Yb and 25(1) μK for ^{174}Yb, which are slightly higher than in previous studies [40, 44] due to an intensity imbalance in the cooling beams. The atom survival probabilities after imaging are 98.7% for ^{171}Yb and 98.8% for ^{174}Yb. Since the isotope shifts of the relevant transitions between ^{171}Yb and ^{174}Yb are on the order of 1 GHz, significantly larger than the natural linewidths of the $^1S_0 - {}^1P_1$ (29 MHz) and $^1S_0 - {}^3P_1$ (182 kHz) transitions [33, 34] (Fig. 6.1c), the probe and cooling beams for one isotope do not induce atom loss in the other. Consequently, the survival probabilities are not degraded compared to those in single-isotope experiments.

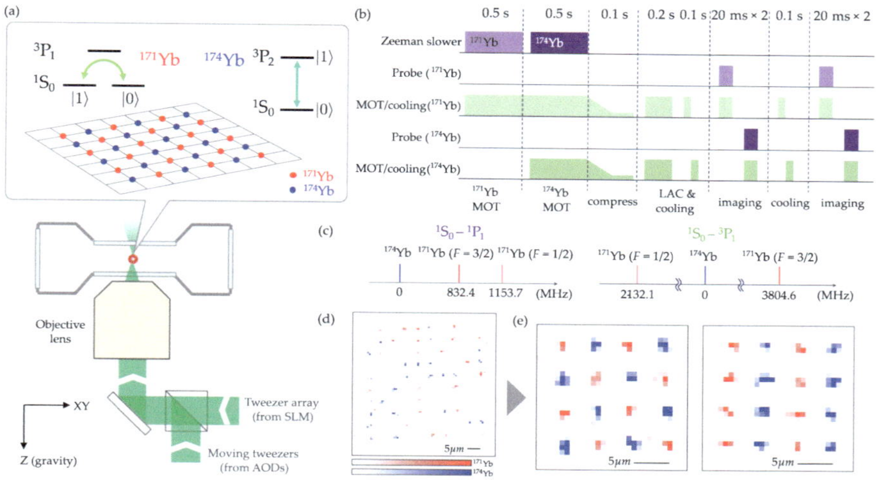

Fig. 6.1 Overview of dual-Ytterbium atom array. **a** The vacuum chamber and the tweezer beam path. The tweezer array beam and moving tweezers are combined with a polarized beam splitter and focused at the center of the chamber through an objective lens with NA = 0.6. The fermionic isotope ^{171}Yb offers a nuclear spin qubit with a long coherence time and fast qubit control. The qubit states are encoded in the ground state. The bosonic isotope ^{174}Yb provides an optical clock qubit encoded in the ground state and a metastable excited state, which has the capability of non-destructive qubit readout. **b** Experimental sequence. First, the dual MOT of ^{171}Yb and ^{174}Yb is prepared by switching the frequency of the Zeeman slower. Then, atoms are loaded into the tweezers by compressing the dual-MOT. After LAC and cooling for each isotope, the atoms are sequentially imaged by a 399 nm probe and 556 nm cooling beams. **c** Isotope shifts of the $^1S_0 \leftrightarrow {}^1P_1$ ($\lambda = 399\,\text{nm}$, $\Gamma/(2\pi) = 29\,\text{MHz}$) [33] and $^1S_0 \leftrightarrow {}^3P_1$ ($\lambda = 556\,\text{nm}$, $\Gamma/(2\pi) = 182\,\text{kHz}$) [34] transitions. We use the $F = 3/2$ state of the $^1P_1$ and $^3P_1$ states for imaging and cooling for ^{171}Yb. **d** Combined single-shot image of dual-Yb array. The optical tweezer array is 10×10 with $5\,\mu\text{m}$ intervals. The red and blue dots represent single ^{171}Yb and ^{174}Yb atoms, respectively. **e** Rearranged dual-Yb atom array in checkerboard (left) and stripe (right) patterns. The atoms are moved between adjacent rows or columns to avoid the filled sites. Modified from my previously published work [1]

6.2.1 Dual Atom Rearrangement

We use 532 nm moving tweezers generated with a pair of acousto-optic deflectors (AODs) to rearrange the atoms into an arbitrary pattern, as shown in Fig. 6.1e. Since the 532 nm tweezers can trap both isotopes and the atoms are randomly loaded in the optical tweezer array, there can be misfilled sites where the ^{174}Yb atoms fill the site designated for ^{171}Yb atoms and vice versa. Additionally, we must prevent the atoms from passing through the correctly-filled sites during the transport of the atoms. These difficulties complicate the rearrangement procedure compared with the single-isotope case. Previous research on a dual-isotope atom array has developed protocols that move the atoms along trajectories on the array grid after removing misplaced atoms that may block the paths [47, 48]. Instead, we adopt trajectories that pass between the rows and columns of the array and extend the parallel sort-and-compression algorithm [49], originally designed for single-isotope cases, to apply to a dual-isotope atom array.

Our protocol involves three steps for each isotope: "classification," "presorting," and "main sorting" (Fig. 6.2). During the classification step, isotope 2 atoms occupying sites designated for isotope 1 are relocated to either correct or non-target sites. In the presorting step, isotope 1 atoms are redistributed between columns to ensure that each column has enough atoms to fill the designated sites for isotope 1. Excess atoms, if any, are removed from the array during this step. Finally, in the main sorting step, isotope 1 atoms are transported to their designated sites column by column, either individually or simultaneously when possible. After completing the rearrangement of isotope 1, the same procedure is repeated for isotope 2. If the number of isotope 1 atoms in each column already matches the number of designated sites, the classification step for isotope 1 is skipped, and the presorting step for isotope 2 begins

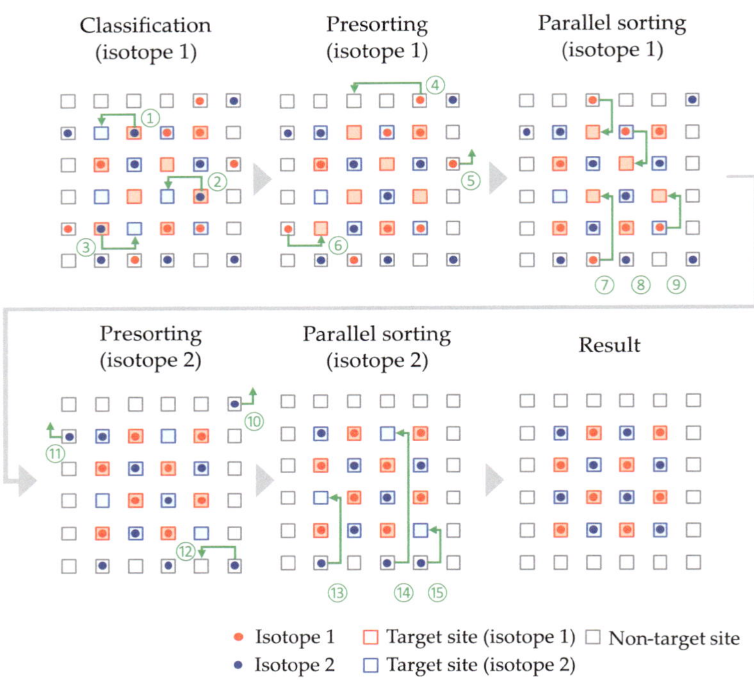

Fig. 6.2 Rearrangement protocol of a dual-isotope atom array. The protocol consists of classification, presorting, and main sorting steps for each isotope. In the classification step, the atoms that occupy incorrect target sites are moved to non-target sites or correct target sites. After pre-sorting to prepare a sufficient number of atoms for each column, the atoms are transported to their target sites. In all steps, the moving tweezers pass between the rows or columns to avoid the occupied sites. Modified from my previously published work [1]

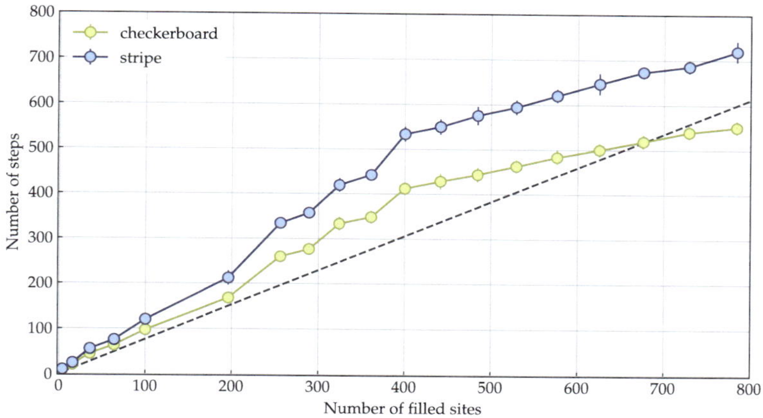

Fig. 6.3 The scaling of the rearrangement protocol. The efficiency is comparable to that in the pioneering work [47] represented as a dashed black line. The scaling slope becomes more gradual as the target array size increases, indicating the efficiency may improve for larger arrays due to parallel sorting. The error bars represent the standard deviations of ten trials. Modified from my previously published work [1]

immediately. Crucially, the moving tweezers in all steps are designed to pass between adjacent rows or columns to avoid interfering with already filled sites.

The scalability of the rearrangement primarily depends on the efficiency of the algorithm. As shown in Fig. 6.3, the scaling of our algorithm is comparable to earlier work on dual-species rearrangement [47] for mid-sized arrays. Interestingly, the gentler scaling slope for larger array sizes indicates that our algorithm may perform even better for large arrays, likely due to the parallel sorting technique demonstrated in [49].

It is worth mentioning that the defect-free array size shown in Fig. 6.1e is currently much smaller than the full 10×10 tweezer array. This limitation arises mainly from nonoptimal transport trajectories, where atoms follow parabolic paths while moving between adjacent rows and columns. In this trajectory, the moving tweezers can approach other filled sites, causing atom loss [11]. We observed that atom loss occurs when the moving tweezer comes within approximately 1μm of filled sites. In our setup, the probability of successfully creating a 4×4 defect-free array from a 10×10 array with a 20% loading probability for each isotope is only about $\sim$1%. However, this success rate could be enhanced through iterative rearrangement, as demonstrated in [49]. By employing a more suitable sweep trajectory, such as the one used in [21], which avoids passing near filled sites, the defect-free array size would primarily depend on the loading probability of each isotope.

6.2.2 *Controlling the Dual Occupancy Probability*

The dual occupancy probability is highly sensitive to the frequency of the LAC beam for ^{174}Yb. By scanning the frequency, we identify the optimal detuning to minimize dual occupancy at $-0.50(2)$ MHz from the $^3P_1$ state resonance in the tweezers (Fig. 6.4). At this detuning, the simultaneous loading probability of both isotopes decreases, indicating the occurrence of heteronuclear single-photon photoassociation (PA). As a result, the loading probabilities for ^{171}Yb and ^{174}Yb are reduced to 21(5)%, with a dual occupancy probability of only 0.2(7)%. To scale up the system size, techniques such as sequential loading with shelving atoms into a metastable state [50] and blue-detuned LAC [40, 51, 52] could improve the loading probabilities, enabling the creation of larger dual-Yb arrays.

The tunability of dual occupancy is a distinctive feature of the dual-Yb array, which is not observed in dual-isotope alkali systems [47]. This difference arises from the disparity in the natural linewidth Γ of the cooling transitions between ytterbium and alkali atoms.

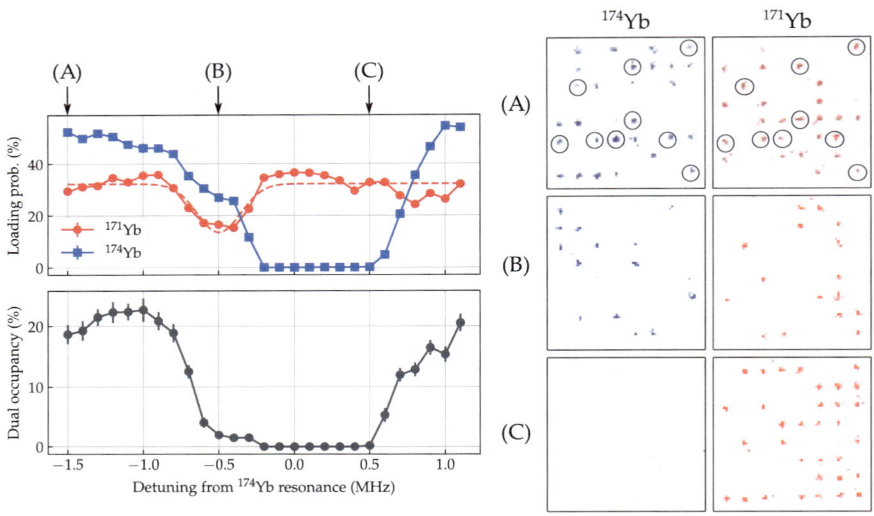

Fig. 6.4 Frequency dependence of dual occupancy probability. Loading probability of each isotope (left top) and dual occupancy (left bottom) as a function of the LAC frequency around the $^3P_1$ resonance of ^{174}Yb. The dip of ^{171}Yb loading probability around -0.5 MHz indicates one-photon heteronuclear PA. The dashed line is a fitting curve by a Gaussian function. The right panels shows typical atom array images at the LAC detuning of -1.5 MHz (A), -0.5 MHz (B), and 0.5 MHz (C), respectively. The black circle represents dual occupancy. Note that the loading probability of each isotope without applying heteronuclear PA fluctuates in the range of $40 \pm 10\%$, depending on the experimental conditions. The error bars represent the standard error of the mean. Modified from my previously published work [1]

The interatomic potential of heteronuclear molecules near the dissociation limit is dominated by the van der Waals interaction, expressed as $V(R) \propto -C_6/R^6$. Using the LeRoy-Bernstein formula [53], the derivative of the binding energy $E(v)$ with respect to the vibrational level v is given by:

$$\frac{dE(v)}{dv} \propto C_6^{-1/6}(D - E(v))^{2/3}, \tag{6.1}$$

where D is the dissociation energy. Since the detuning of the cooling beams, which induces light-assisted collisions or photo-association, is on the order of several Γ, $D - E(v)$ can be approximated by Γ. Additionally, as the C_6 coefficient is derived from the second-order perturbation of the dipole-dipole interaction with strength d^2, we have $C_6 \propto |d^2|^2 = \Gamma^2$. Substituting this into the equation, the scaling of the binding energy becomes:

$$\frac{dE(v)}{dv} \propto (\Gamma^2)^{-1/6}\Gamma^{2/3} = \Gamma^{1/3}. \tag{6.2}$$

From this, the number of weakly-bound states per Γ can be estimated as $\Gamma/(dE/dv) = \Gamma^{2/3}$.

Given that the linewidth ratio between the ${}^1S_0$–${}^3P_1$ transition of ytterbium and the D_2 transition of rubidium is approximately 1 : 30, the number of bound states per Γ for alkali atoms is about 10 times greater than that for ytterbium. Consequently, the cooling beams for alkali atoms are more likely to overlap with photo-association resonances, naturally inducing heteronuclear light-assisted collisions. In contrast, the photo-association resonances of ytterbium are better resolved relative to the linewidth of the ${}^1S_0$–${}^3P_1$ transition, as discussed later. It is worth noting that dual occupancy is inherently avoided in dual-species alkali systems by employing two-color tweezer setups [35, 54].

6.2.3 Observation of Heteronuclear Photoassociation Resonances

Additionally, we explore other photoassociation (PA) resonances and identify four resonances at −268(2), −281(1), −698.8(8), and −716.4(5) MHz (Fig. 6.5). To confirm that these lines correspond to heteronuclear PA resonances, we perform a theoretical calculation of the PA resonance positions.

To compute the PA resonances of the excited ^{171}Yb^{174}Yb molecule, which asymptotically connects to the ${}^1S_0$–${}^3P_1$ state, we begin with the interaction Hamiltonian under the Born-Oppenheimer approximation [55], expressed as:

$$\mathcal{H}_{\text{int}} = \frac{\boldsymbol{d}_{171} \cdot \boldsymbol{d}_{174} - 3d_{z,171}d_{z,174}}{4\pi\epsilon_0 R^3} + Ah\boldsymbol{I}_{171} \cdot \boldsymbol{J}_{171}. \tag{6.3}$$

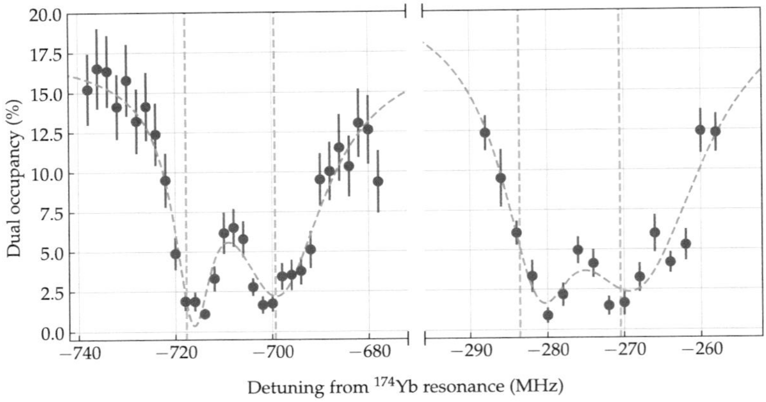

Fig. 6.5 Heteronuclear PA resonances of deeper bound states around −710 MHz and −280 MHz. The data are fitted by a double Lorentzian function. The yellow dashed lines represent the theoretical PA resonances. The error bars represent the standard error of the mean. Modified from my previously published work [1]

Here, $\boldsymbol{d}$ and d_z represent the dipole operator and its projection onto the interatomic axis for each atom, ϵ_0 is the vacuum permittivity, R is the interatomic distance, h is Planck's constant, A is the hyperfine coupling constant of ^{171}Yb, and $\boldsymbol{I}$ and $\boldsymbol{J}$ are the nuclear spin and electronic angular momentum operators of ^{171}Yb, respectively. The basis states used for diagonalization are $|G_i\rangle_{171} \otimes |E_j\rangle_{174}$ and $|E_i\rangle_{171} \otimes |G\rangle_{174}$, where $|G_i\rangle_{171}$ $(i = 1, 2)$ and $|E_i\rangle_{171}$ $(i = 1$–$6)$ denote the ground and excited ^{3}P$_1$ states of ^{171}Yb, and $|G\rangle_{174}$ and $|E_j\rangle_{174}$ $(j = 1$–$3)$ represent the ground and excited ^{3}P$_1$ states of ^{174}Yb, respectively. By diagonalizing the Hamiltonian (Eq. (6.3)), we derive six interaction potentials, as depicted in Fig. 6.6. The retardation effect on the resonant dipole interaction [56] is included in this calculation. Among these, potentials Nos. 1–3 are attractive and can support weakly-bound heteronuclear molecules.

Next, we incorporate a centrifugal potential, van der Waals potential, and a short-range repulsive potential into the derived potential $\langle \mathcal{H}_{\text{int}} \rangle$ to construct a model potential [57]:

$$V(R) = \langle \mathcal{H}_{\text{int}} \rangle + \hbar^2 \frac{T_e(T_e + 1) + \langle \boldsymbol{F}^2 \rangle - 2\Omega^2}{2\mu R^2} - \frac{C_6}{R^6} + \frac{C_{12}}{R^{12}}, \tag{6.4}$$

where $\hbar = h/(2\pi)$ is the reduced Planck constant, μ is the reduced mass of ^{171}Yb and ^{174}Yb, T_e and Ω are the total electronic angular momentum and its projection onto the molecular axis, and C_6 and C_{12} are the coefficients for the van der Waals and short-range repulsive potentials, respectively. The term $\boldsymbol{F} = \boldsymbol{I}_{171} + \boldsymbol{J}_{171} + \boldsymbol{J}_{174}$ represents the sum of the nuclear spin of ^{171}Yb and the electronic angular momenta of ^{171}Yb and ^{174}Yb. Finally, we solve the Schrödinger equation for the model potential (Eq. (6.4)) using the Numerov method [58] to obtain the eigenenergies. The parameters used in this calculation are summarized in Table 6.1.

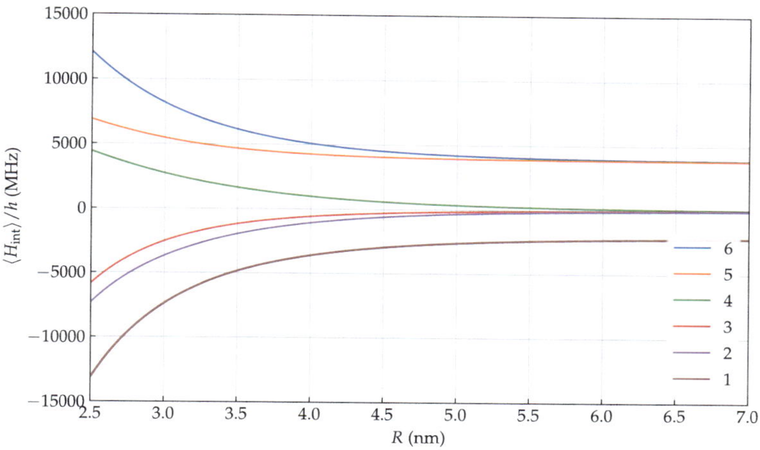

Fig. 6.6 Interaction potentials between ^{171}Yb and ^{174}Yb obtained from Eq. (6.3). The potentials are plotted with respect to the $^3P_1$ state of ^{174}Yb. Potential Nos. 2 and 3 asymptotically connect to the $|{}^1S_0\rangle_{171}\,|{}^3P_1\rangle_{174}$ atomic state in the dissociation limit. Modified from my previously published work [1]

Table 6.1 Parameters for PA resonance calculation. E_h is Hartree energy and a_0 is Bohr radius

Parameter	Value
A	3957.781 MHz [34]
$d^2/(4\pi\epsilon_0)$	0.08744313 $E_h a_0^3$ [59]
C_6	2405.364747 $E_h a_0^6$ [59]
C_{12}	9.319×10^8 $E_h a_0^{12}$ (fitting)

Table 6.2 shows the resonance frequencies of the PA below 1 GHz for the potential No. 2, in which the molecular states have $T_e = 1/2,\ 3/2$ and $|\Omega| = 1/2$. The calculated PA resonance frequencies agree with the experimental values with a few MHz precision.

These findings highlight the potential to utilize these PA lines for techniques such as stimulated Raman adiabatic passage [60, 61] or two-photon PA [62, 63] to create ground-state heteronuclear molecules in optical tweezers [63–69]. Such molecules offer multiple degrees of freedom, including rotational and vibrational levels, as well as nuclear spin states decoupled from hyperfine interactions. The presence of nuclear spin states with long coherence times [70, 71] enables the realization of proposals leveraging molecular degrees of freedom for quantum computation, potentially enhancing system scalability [72–74]. While polar molecules formed from alkali atoms also feature nuclear spin states in the $^1\Sigma$ electronic ground state, their preparation is typically more technically challenging [75]. In contrast, the dual-Yb system provides a relatively straightforward approach [76–78] for producing heteronuclear molecules with nuclear spin states.

Table 6.2 PA resonances below 1 GHz for the potential No. 2

T_e	Calculation (MHz)	Experiment (MHz)
1/2	−0.05	−0.50(2)
3/2	−55.3	–
1/2	−62.9	–
3/2	−270.5	−268(2)
1/2	−283.6	−281(1)
3/2	−699.4	−698.8(8)
1/2	−717.8	−716.4(5)

Additionally, the cooling beams used during imaging do not induce heteronuclear PA, as they are detuned by approximately -3Γ from the PA resonance, as shown in panel (A) of Fig. 6.4. This enables various operations to be performed while maintaining dual occupancy sites, such as spin initialization of ^{171}Yb, motional cooling via Raman sideband cooling [40, 79] for ^{171}Yb, and clock transition cooling [80] for ^{174}Yb, provided the operation lasers are detuned from the PA resonances. Furthermore, it is possible to image atoms dissociated from a molecule after specific operations, allowing the internal state of the molecule to be detected without losing the atoms. These capabilities not only improve the fidelity of molecule creation but also simplify molecular experiments, making them more efficient and accessible.

6.3 Crosstalk Measurement

In the dual-Yb atom array, two types of qubits can be utilized: nuclear spin qubits and optical qubits. The nuclear spin qubit is encoded in the ground state of ^{171}Yb as $|0\rangle = \left|{}^1\mathrm{S}_0, m_F = 1/2\right\rangle$ and $|1\rangle = \left|{}^1\mathrm{S}_0, m_F = -1/2\right\rangle$, offering second-order coherence time and MHz-order fast control ([40, 41], and see Sect. 5.1). The optical qubit, encoded in the ground state and the $m_J = 0$ state of an excited metastable state of ^{174}Yb, such as $^3\mathrm{P}_0$, $^3\mathrm{P}_2$, and $4f^{13}5d6s^2$ $(J = 2)$ [81, 82], can be controlled via M2-transition [83] or magnetic-field-induced E1 transition [84, 85] and imaged non-destructively, as the metastable state naturally acts as a dark state for the probe transition, as demonstrated in recent work using the $^3\mathrm{P}_0$ state [32].

We examine the crosstalk between these qubits, focusing on the ability to image ^{174}Yb while preserving the coherence of the nuclear spin qubits. Since the resonant frequencies of the probe and cooling light for ^{174}Yb are well-separated from those for ^{171}Yb by several GHz (Fig. 6.1c), the photon scattering rates of ^{171}Yb during ^{174}Yb imaging are calculated to be approximately 30 Hz for 399 nm light and below 0.1 Hz for 556 nm light, which are negligible for our imaging conditions with a 20 ms exposure time. Notably, ^{174}Yb is the optimal isotope for ^{171}Yb in terms of isotope shifts to minimize crosstalk [33, 34].

To quantify this, we evaluate the coherence properties of the nuclear spin qubit using Ramsey and Hahn-echo sequences under irradiation of imaging light for ^{174}Yb. In these experiments, two imaging methods are considered: "399 nm imaging" using a 399 nm probe beam and 556 nm cooling beams, and "556 nm imaging" using a single 556 nm probe beam.

The experiment begins by preparing the dual-Yb atom array and initializing the nuclear spin qubit via optical pumping on the $|1\rangle \rightarrow |^3P_1, F = 3/2, m_F = 1/2\rangle$ transition. A circularly-polarized single 556 nm beam in the horizontal plane is then used to rotate the qubits via single-beam Raman transitions [40]. The qubit control beam has an intensity of $2.4 \times 10^5\ I_{s,556}$ and a detuning of -1.0 GHz (-3.1 GHz) from the resonance of $|^3P_1, F = 1/2\rangle$ of ^{171}Yb ($|^3P_1\rangle$ of ^{174}Yb), resulting in a Rabi frequency of $2\pi \times 208\ kHz$. After a $\pi/2$-pulse and a waiting time of $T + \Delta T$, another $\pi/2$-pulse is applied to map the relative phase of the qubit states to their populations. The additional waiting time ΔT is used to scan the relative phase. In the Hahn-echo sequence, an additional π-pulse is applied after a waiting time of $T/2$ from the first $\pi/2$-pulse. Finally, atoms in the $|0\rangle$ state are removed via the $|0\rangle \leftrightarrow |^3P_1, F = 3/2, m_F = 3/2\rangle$ transition, and the remaining atoms are imaged. By scanning ΔT, we obtain Ramsey (Hahn-echo) signals and compare their contrasts with and without imaging light for ^{174}Yb during the holding time T to assess its impact.

6.3.1 *Hahn-Echo Sequence*

We first present the results of the Hahn-echo experiment using our current 399 nm imaging method. During the holding time T, with a trap depth of 0.9 mK, we apply a resonant 399 nm probe beam with a radius of 1.3 mm at an intensity of $1 \times 10^{-3}\ I_{s,399}$, as well as 556 nm cooling beams with radii of 4 mm (vertical) and 12 mm (horizontal) at a total intensity of $10\ I_{s,556}$, which are detuned by -5.5Γ from the resonance. Figure 6.7 shows the Hahn-echo signals for holding time of $T = 20$ ms, which is typical in our current setup, with and without 399 nm imaging. The oscillation frequency is determined by the Zeeman splitting of the nuclear spin states with a magnetic field of 2.5 G applied during the holding time. The contrast ratio between the two conditions is 0.991(18), indicating that the coherence is maintained within the uncertainty.

6.3.1.1 Imaging Performance

The detection fidelity of $|^1S_0\rangle = |0\rangle$ of ^{174}Yb is a crucial parameter. To assess the imaging performance, particularly the discrimination fidelity of the states, we utilize two approaches: the conventional fitting method and a model-free method [12].

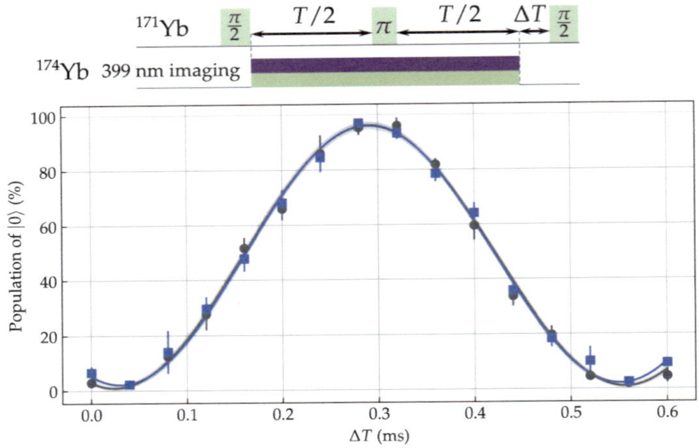

Fig. 6.7 Influence on the coherence of the nuclear spin qubit from the imaging light for ^{174}Yb in the Hahn-echo sequence. Hahn-echo signal for holding time of $T = 20$ ms with (blue square) and without (black circle) 399 nm probe and 556 nm cooling beams. The solid lines are the fitting curves by a sinusoidal function. The contrast ratio is 0.991(18), highlighting the possibility of the 399 nm imaging without disturbing the coherence of the nuclear spin qubits. The error bars represent the standard error of the mean. Modified from my previously published work [1]

The first method involves fitting the histogram of detected photon counts during imaging, as shown in Fig. 6.8. The fitting function is expressed as:

$$P(x) = (1 - F)P_D(x) + FP_B(x), \tag{6.5}$$

where F represents the atom loading probability, and $P_D(x)$ and $P_B(x)$ correspond to the distributions for the dark and bright peaks of the histogram:

$$\begin{aligned} P_D(x) &= \frac{1}{\sqrt{2\pi}\sigma_D} \exp\left(-\frac{(x-\mu_D)^2}{2\sigma_D^2}\right) + a\exp(-bx), \\ P_B(x) &= \frac{1}{\sqrt{2\pi}\sigma_B} \exp\left(-\frac{(x-\mu_B)^2}{2\sigma_B^2}\right)\left(1 + \text{erf}\left(c\frac{x-\mu_B}{\sqrt{2}\sigma_B}\right)\right), \end{aligned} \tag{6.6}$$

with fitting parameters $\sigma_D,\ \mu_D,\ a,\ b,\ \sigma_B,\ \mu_B,\ c$. The $P_D(x)$ combines a Gaussian and an exponential distribution to account for readout and amplification noise from the EMCCD camera. While the amplification process is more accurately described by the Erlang distribution $P(x) = x^{n-1}e^{-x/g}/(g^n(n-1)!)$, where n is the photo-electron number and g is the gain [86], it simplifies to an exponential distribution for the dark distribution when $n \leq 1$. For the bright peak, a skewed Gaussian distribution is used for computational simplicity, as in prior work [19].

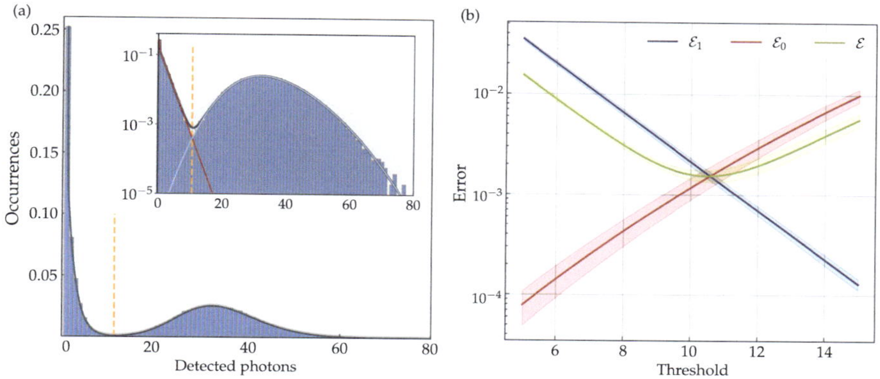

Fig. 6.8 Fitting to the histogram of detected photons for imaging performance evaluation. **a** The histogram of the number of detected photons by 399 nm imaging with an exposure time of 20 ms. The black, red, and light blue curves represent the fitting function (Eq. 6.7), $P_D(x)$, and $P_B(x)$, respectively. The dashed orange is the threshold that maximizes the discrimination fidelity at 0.9985(2). **b** Imaging infidelities as a function of the threshold. The shaded areas in (**a**) and (**b**) are the 1σ confidence intervals. Modified from my previously published work [1]

Figure 6.8 illustrates the fitting results and the corresponding infidelities, calculated as:

$$\begin{aligned} \mathcal{E} &= (1-F)\mathcal{E}_0 + F\mathcal{E}_1, \\ &= (1-F)\int_{x_{\mathrm{th}}}^{\infty} P_D(x)dx + F\int_0^{x_{\mathrm{th}}} P_B(x)dx, \end{aligned} \tag{6.7}$$

where x_{th} is the threshold for determining the presence of atoms, and $\mathcal{E}_0$ and $\mathcal{E}_1$ represent false-negative and false-positive errors, respectively. The optimal fidelity is determined to be 0.9985(2).

To evaluate the fidelity without bias, we also use a model-free method as described in [12]. In this approach, three consecutive images are captured under identical conditions during the cross-talk measurement using 399 nm imaging. For three images taken in a single sequence, the probabilities of each event $\tilde{P}_{x_1x_2x_3}$, where $x_i \in \{0, 1\}$ indicates whether a single atom is trapped (1) or not (0) in the i-th shot, are given by:

$$\begin{aligned} \tilde{P}_{111} &= S^2F, \\ \tilde{P}_{110} &= (1-S)SF, \\ \tilde{P}_{100} &= (1-S)F, \\ \tilde{P}_{000} &= 1-F. \end{aligned} \tag{6.8}$$

Here, S is the survival probability after imaging, and F is the loading (filling) probability. Given a threshold for determining atom presence, the experimental probability of each event is calculated as:

$$P_{y_1y_2y_3} = \sum_{x_1x_2x_3} P(y_1|x_1)P(y_2|x_2)P(y_3|x_3)\tilde{P}_{x_1x_2x_3}, \tag{6.9}$$

where the conditional probabilities $P(y|x)$ are derived from the discrimination fidelities $\mathcal{F}_0$ and $\mathcal{F}_1$ as $P(1|1) = \mathcal{F}_1$, $P(0|1) = 1 - \mathcal{F}_1$, $P(1|0) = 1 - \mathcal{F}_0$, and $P(0|0) = \mathcal{F}_0$. These equations expand as:

$$\begin{aligned}
P_{111} &= (1-\mathcal{F}_0)^3\tilde{P}_{000} + \mathcal{F}_1(1-\mathcal{F}_0)^2\tilde{P}_{100} + \mathcal{F}_1^2(1-\mathcal{F}_0)\tilde{P}_{110} + \mathcal{F}_1^3\tilde{P}_{111},\\
P_{110} &= (1-\mathcal{F}_0)^2\mathcal{F}_0\tilde{P}_{000} + \mathcal{F}_1(1-\mathcal{F}_0)\mathcal{F}_0\tilde{P}_{100} + \mathcal{F}_1^2\mathcal{F}_0\tilde{P}_{110} + \mathcal{F}_1^2(1-\mathcal{F}_1)\tilde{P}_{111},\\
P_{100} &= (1-\mathcal{F}_0)\mathcal{F}_0^2\tilde{P}_{000} + \mathcal{F}_1\mathcal{F}_0^2\tilde{P}_{100} + \mathcal{F}_1(1-\mathcal{F}_1)\mathcal{F}_0\tilde{P}_{110} + \mathcal{F}_1(1-\mathcal{F}_1)^2\tilde{P}_{111},\\
P_{000} &= \mathcal{F}_0^3\tilde{P}_{000} + (1-\mathcal{F}_1)\mathcal{F}_0^2\tilde{P}_{100} + (1-\mathcal{F}_1)^2\mathcal{F}_0\tilde{P}_{110} + (1-\mathcal{F}_1)^3\tilde{P}_{111}.
\end{aligned} \tag{6.10}$$

By adjusting the parameters F, S, $\mathcal{F}_0$, and $\mathcal{F}_1$ to align theoretical and experimental probabilities, the discrimination fidelity is determined as $\mathcal{F} = (1-F)\mathcal{F}_0 + F\mathcal{F}_1$. This process is repeated for various thresholds to find the optimal fidelity that minimizes infidelity. Using this method, we achieve a discrimination fidelity of $\mathcal{F} = 0.9992$ with a survival probability $S = 0.988$ under a 20 ms exposure time and a loading probability $F = 0.561$, as shown in Fig. 6.9.

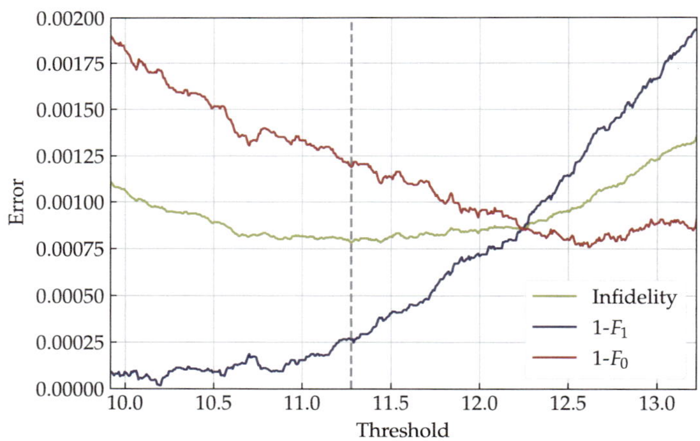

Fig. 6.9 Imaging fidelity evaluation by model-free method. The best fidelity is found to be 0.9992 with a survival probability of 0.988 under a 20 ms exposure time. The loading probability is found to be 0.561 in this measurement. The dashed line indicates the threshold that minimize the discrimination infidelity. Modified from my previously published work [1]

6.3.1.2 Long Exposure Time Region

To investigate the impact on coherence over extended exposure times, we increase the holding time up to 1500 ms. Figure 6.10 shows the decay of the Hahn-echo contrast as a function of exposure time. The coherence time T_2 is extracted by fitting the data to the function $f(T) = A\exp(-(T/T_2)^n)$, where A, T_2, and n are fitting parameters. The fitting results are presented in Table 6.3. For 399 nm imaging, the coherence time is measured to be 0.67(5) s, reduced due to scattering from the probe and cooling beams. However, as shown in Fig. 6.7, high-fidelity imaging is still achievable while preserving coherence. Additionally, considering the numerical aperture (NA) of our objective lens, the photon collection efficiency could potentially be improved by a factor of three. With such an optimized imaging system, the exposure time could be reduced by one-third, enabling a data qubit coherence retention of 99.7%. It is worth noting that the coherence without imaging is currently limited by scattering from the tweezer at a trapping depth of 0.9 mK. This limitation could be mitigated by reducing the trap depth at the data qubit sites while maintaining a deeper trap at the ancilla qubit sites.

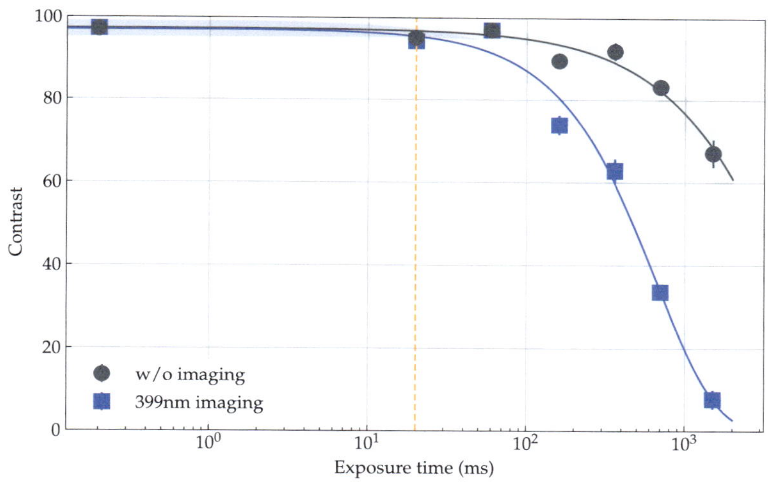

Fig. 6.10 Decay curves of the Hahn-echo contrasts with and without exposure to imaging light for ^{174}Yb. The solid lines are fitting curves by $f(T) = A\exp(-(T/T_2)^n)$. The fitting results are summarized in Table 6.3. Although the coherence under the 399 nm imaging decreases faster than that of without imaging, the contrast of the Hahn-echo signal is almost the same for our typical exposure time of 20 ms, represented as an orange dashed line. The error bars and shaded regions represent the standard error of the mean and 1σ confidence intervals of the fit, respectively. Modified from my previously published work [1]

Table 6.3 Fitting results of the coherence curves in the Hahn-echo sequence

	A	T_2 (s)	n
Without imaging	0.97(2)	4.3(2.3)	1.0(3)
399 nm imaging	0.97(2)	0.67(5)	1.2(2)

6.3.2 Ramsey Sequence

To explore potential improvements, we assess the impact on coherence using the Ramsey sequence with 556 nm imaging, which has demonstrated high effectiveness for ^{174}Yb trapped in 532 nm tweezers [44]. The trap depth remains at 0.9 mK, consistent with the Hahn-echo sequence. During the holding time, we apply a resonant 556 nm probe beam for ^{174}Yb, with a radius of 1.4 mm and an intensity of 3 $I_{s,556}$, as used in prior studies [44]. The probe beam propagates diagonally at a 16° angle from the horizontal plane and is horizontally polarized. Figure 6.11 shows the Ramsey contrast trends for holding times up to 160 ms. While the contrast under 399 nm imaging decreases rapidly, the coherence with 556 nm imaging remains nearly identical to that without imaging. We again note that the observed coherence reduction is

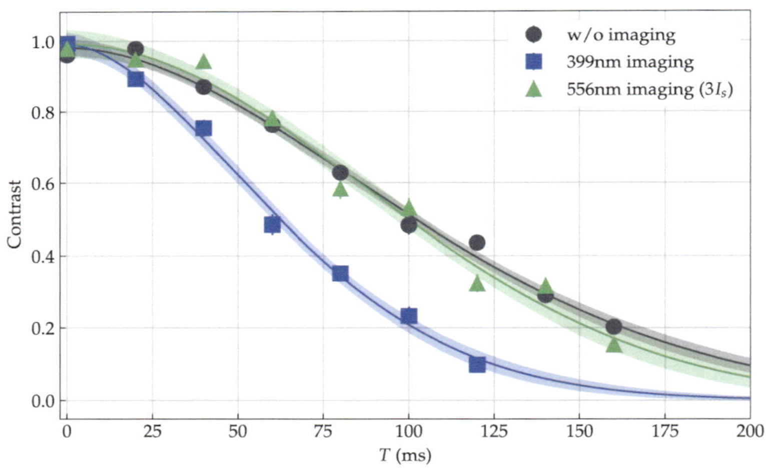

Fig. 6.11 Influence on the coherence of the nuclear spin qubit from the imaging light for ^{174}Yb in the Ramsey sequence. Decay curves of the Ramsey contrast are shown as a function of exposure time. The solid lines are the fitting curves by $f(T) = A\exp(-(T/T_2^*)^n)$, where A, T_2^*, and n are fitting parameters. While the Ramsey contrast under 399 nm imaging decays faster than that without imaging, the contrast with 556 nm probe light with an intensity of 3 $I_{s,556}$ (green triangle) remains the same as that of without irradiating a probe beam, suggesting that 556 nm imaging is a possible future improvement of low cross-talk measurements. The error bars and shaded regions represent the standard error of the mean and 1σ confidence intervals of the fit, respectively. Modified from my previously published work [1]

Table 6.4 Fitting results of the coherence curves in the Ramsey sequence

	A	T_2^* (ms)	n
Without imaging	0.98(2)	130(3)	2.0(2)
556 nm imaging	0.98(2)	122(5)	2.3(3)
399 nm imaging	0.99(2)	77(2)	1.8(1)

due to the 0.9 mK tweezer potential and does not represent a fundamental limitation. The fitting results, using the same function as in the Hahn-echo experiment, are summarized in Table 6.4. These findings highlight the potential of 556 nm imaging for low cross-talk measurements. Based on prior results [44], the discrimination fidelity is expected to exceed 0.9999 with a 99% survival probability for a 30 ms exposure time, with negligible impact on data qubit coherence. Since our current optical setup does not support 556 nm imaging, we plan to evaluate its performance in the dual-Yb system in future work. This will provide a comprehensive assessment of its potential to enhance low cross-talk measurements and overall system performance.

6.4 Discussion

The cross-talk measurement using our current 399 nm imaging, as shown in Fig. 6.7, demonstrates that the dual-Yb system enables low cross-talk measurements with high-fidelity detection of ancilla qubits. The current imaging performance meets the requirements for implementing surface codes in neutral atom arrays [87]. Furthermore, since surface codes are generally more tolerant of readout errors than coherent errors [88], our imaging performance, especially with further optimization, is likely acceptable for surface code implementation. The coherence loss during imaging is $<1\%$, which is comparable to or better than previous studies employing the omg-architecture [32], spectral isolation via local light shifts [31], or shelving techniques [30] (Table 6.5). A key advantage of the dual-Yb system is its simplicity in performing mid-circuit measurements, unlike other approaches, such as spatial isolation methods [21, 29], which require technically complex local operations.

As is shown in Fig. 6.7, the crosstalk induced by the 399 nm imaging with our current parameters is only 0.9%. However, when comparing the dual-Yb system with the dual-alkali system [36], the behavior of the contrast decay curves with and without imaging in the long exposure time region differs from that observed in the dual-alkali system. This is due to the limited frequency isolation, which is characterized by the ratio between the isotope shift $|\omega - \omega'|$ and the natural linewidth Γ of the transition, with $|\omega_{^{171}\mathrm{Yb}} - \omega_{^{174}\mathrm{Yb}}|/\Gamma_{399} \sim 30$ for the $^1S_0 \leftrightarrow {}^1P_1$ transition of ytterbium, much smaller than $|\omega_{\mathrm{Rb}} - \omega_{\mathrm{Cs}}|/\Gamma \sim 10^6$ for the dual-alkali system. However, as shown in

Table 6.5 Comparison of low cross-talk measurement schemes

	Readout time (ms)	Imaging fidelity	Figure of merit
Shelving [30]	4	94.9(8)% ($\vert 0\rangle$) 95.3(1.1)% ($\vert 1\rangle$)	$\mathcal{F} = 97.0(5)\%$[a]
Local light shift [31]	5	∼99.9%	97.5(1)%($\vert 0\rangle$ imaging)[b] 98.7(8)%($\vert 1\rangle$ imaging)[b]
Omg-architecture [32]	3.5	99.8%	$F_{p,DQ} = 97.2(5)\%$ [c]
Dual-alkali [36]	15	98.9%	$(A, \tau, n) = (0.97(2),$ 650(20) ms, 1.71(9))[d] (w/o readout) $(A, \tau, n) = (0.935(2),$ 678(13) ms, 1.82(7))[d] (w/ readout)
Dual-Yb	20	99.92% (98.8% survival)	99.1(1.8)%[e]

[a] They estimated the process fidelity $\mathcal{F}$ of the shelving and echo by a Ramsey-type experiment with a measurement of ancilla qubits

[b] They compared the coherence contrast of the data qubits with and without imaging of the ancilla qubits during the holding time of the Ramsey sequence

[c] They estimated the process fidelity $F_{p,DQ}$ of mid-circuit measurement by quantum process tomography

[d] They fitted the contrast decay curves of the data qubits as a function of readout time using the function $f(t) = A\exp(-(t/\tau)^n) + B$ and compared the τ for cases with and without readout of ancilla qubits

[e] German We compared the coherence contrast of the data qubits with and without imaging of the ancilla qubits during the hold time of the Hahn-echo sequence

Fig. 6.5, 556 nm imaging holds the potential to achieve comparably low cross-talk measurements as the dual-alkali system, owing to its large frequency isolation of $|\omega_{^{171}\mathrm{Yb}} - \omega_{^{174}\mathrm{Yb}}|/\Gamma_{556} \sim 10^4$.

It is worth noting the key distinction between alkali and Yb systems lies in the types of qubits utilized. In alkali systems, hyperfine state qubits with millisecond-scale coherence times (without dynamical decoupling) are employed for both data and ancilla qubits. Qubit detection in these systems often relies on state-selective atom loss prior to imaging, necessitating more frequent atom replenishment. While non-destructive imaging has been demonstrated in alkali systems using techniques like shelving [30] or state transfer to a stretched state [19, 89], these methods may introduce finite infidelity. In contrast, the dual-Yb system provides a nuclear spin qubit as the data qubit, offering inherently long second-order coherence times, and an optical qubit as the ancilla qubit, which naturally supports non-destructive readout without requiring state-selective loss. These features make the dual-Yb system particularly advantageous for implementing mid-circuit measurements in quantum error correction protocols.

6.5 Conclusion

In conclusion, we have demonstrated the realization of a dual-Yb atom array comprising fermionic ^{171}Yb and bosonic ^{174}Yb. This hybrid system features a nuclear spin qubit as a data qubit with long coherence times and an optical clock qubit as an ancilla qubit, enabling non-destructive qubit state readout. We observed tunable dual occupancy rates ranging from near zero to several tens of percent, highlighting the system's versatility for experiments involving heteronuclear molecules in optical tweezers [90–94].

We evaluated the coherence of the nuclear spin qubits under exposure to the imaging beams for ^{174}Yb. For 399 nm imaging with a 20 ms exposure time, coherence was preserved at the 99.1(1.8) % level, with a discrimination fidelity of 99.92 % and a survival probability of 98.8% for ^{174}Yb. While the current imaging performance is slightly below the state-of-the-art for divalent atom arrays [22, 32], the coherence retention during ancilla readout is comparable to previous works [30–32], even without requiring complex local operations. Additionally, we confirmed that 556 nm imaging with 3 $I_{s,556}$, as used in prior studies [44], has no adverse effect on coherence and offers high-fidelity, low-loss imaging, making it a strong candidate for mid-circuit measurements. Combined with further optimization of the imaging system, we anticipate that the dual-Yb system will be fully capable of mid-circuit operations.

Looking ahead, we aim to incorporate coherent control of optical qubits within the dual-Yb system. The ^{3}P$_2$ and $4f^{13}5d6s^2$ $(J = 2)$ states, along with the ^{3}P$_0$ state, are promising candidates for encoding optical qubits due to their insensitivity to magnetic field fluctuations [95] and compatibility with magic-wavelength tweezers that mitigate tensor light shifts. Single-qubit operations on these optical qubits will utilize π-pulses for X gates and phase shift of the excitation laser for Z gates [17]. Additionally, the fine-structure qubit recently demonstrated in strontium systems [96, 97] presents an exciting avenue, particularly for its potential in fast and precise control.

To generate entanglement between nuclear spin qubits and clock transition qubits, time-optimal two-qubit gates mediated by Rydberg interactions [98] can be employed. This requires exploring the Rydberg interaction between isotopes. Recent advancements in Yb Rydberg spectroscopy [16] and sophisticated software for calculating Rydberg interactions in divalent atoms [99] will facilitate identifying optimal conditions for achieving high-fidelity two-qubit gates, as demonstrated in single-species atom arrays [14–19]. Furthermore, the natural disparity between intra- and inter-isotope Rydberg interactions could enable high-fidelity native multi-qubit gates [100], providing an efficient pathway for quantum computation. These advancements, combined with further improvements and developments in our dual-Yb system, will pave the way for implementing ancilla-based QEC protocols in neutral atom architectures.

References

1. Y. Nakamura, T. Kusano, R. Yokoyama, K. Saito, K. Higashi, N. Ozawa, T. Takano,Y. Takasu, Y. Takahashi, Phys. Rev. X **14**(4), 041062 (2024). https://doi.org/10.1103/PhysRevX.14.041062, https://journals.aps.org/prx/abstract/10.1103/PhysRevX.14.041062
2. D. Gottesman, Stabilizer codes and quantum error correction. Ph.D. Thesis, California Institute of Technology (1997), https://thesis.library.caltech.edu/2900/2/THESIS.pdf
3. Google Quantum AI, Nature (London) **595**(7867), 383 (2021), http://dx.doi.org/10.1038/s41586-021-03588-y
4. Y. Zhao, Y. Ye, H.L. Huang, Y. Zhang, D. Wu, H. Guan, Q. Zhu, Z. Wei, T. He, S. Cao, F. Chen, T.H. Chung, H. Deng, D. Fan, M. Gong, C. Guo, S. Guo, L. Han, N. Li, S. Li, Y. Li, F. Liang, J. Lin, H. Qian, H. Rong, H. Su, L. Sun, S. Wang, Y. Wu, Y. Xu, C. Ying, J. Yu, C. Zha, K. Zhang, Y.H. Huo, C.Y. Lu, C.Z. Peng, X. Zhu, J.W. Pan, Phys. Rev. Lett. **129**(3), 030501 (2022). https://doi.org/10.1103/PhysRevLett.129.030501
5. S. Krinner, N. Lacroix, A. Remm, A. Di Paolo, E. Genois, C. Leroux, C. Hellings, S. Lazar, F. Swiadek, J. Herrmann, G.J. Norris, C.K. Andersen, M. Müller, A. Blais, C. Eichler, A. Wallraff, Nature (London) **605**(7911), 669 (2022). https://doi.org/10.1038/s41586-022-04566-8, https://www.nature.com/articles/s41586-022-04566-8
6. L. Egan, D.M. Debroy, C. Noel, A. Risinger, D. Zhu, D. Biswas, M. Newman, M. Li, K.R. Brown, M. Cetina, C. Monroe, Nature (London) **598**(7880), 281 (2021). https://doi.org/10.1038/s41586-021-03928-y
7. M.P. da Silva, C. Ryan-Anderson, J.M. Bello-Rivas, A. Chernoguzov, J.M. Dreiling, C. Foltz, F. Frachon, J.P. Gaebler, T.M. Gatterman, L. Grans-Samuelsson, D. Hayes, N. Hewitt, J. Johansen, D. Lucchetti, M. Mills, S.A. Moses, B. Neyenhuis, A. Paz, J. Pino, P. Siegfried, J. Strabley, A. Sundaram, D. Tom, S.J. Wernli, M. Zanner, R.P. Stutz, K.M. Svore (2024), arXiv:2404.02280
8. P. Huft, Y. Song, T.M. Graham, K. Jooya, S. Deshpande, C. Fang, M. Kats, M. Saffman, Phys. Rev. A **105**(6), 063111 (2022). https://doi.org/10.1103/PhysRevA.105.063111, https://link.aps.org/doi/10.1103/PhysRevA.105.063111
9. R. Tao, M. Ammenwerth, F. Gyger, I. Bloch, J. Zeiher, Phys. Rev. Lett. **133**(1), 013401 (2024). https://doi.org/10.1103/PhysRevLett.133.013401
10. L. Pause, L. Sturm, M. Mittenbühler, S. Amann, T. Preuschoff, D. Schäffner, M. Schlosser, G. Birkl, Optica **11**(2), 222 (2024). https://doi.org/10.1364/OPTICA.513551, https://opg.optica.org/abstract.cfm?uri=optica-11-2-222
11. M.A. Norcia, H. Kim, W.B. Cairncross, M. Stone, A. Ryou, M. Jaffe, M.O. Brown, K. Barnes, P. Battaglino, T.C. Bohdanowicz, A. Brown, K. Cassella, C.A. Chen, R. Coxe, D. Crow, J. Epstein, C. Griger, E. Halperin, F. Hummel, A.M.W. Jones, J.M. Kindem, J. King, K. Kotru, J. Lauigan, M. Li, M. Lu, E. Megidish, J. Marjanovic, M. McDonald, T. Mittiga, J.A. Muniz, S. Narayanaswami, C. Nishiguchi, T. Paule, K.A. Pawlak, L.S. Peng, K.L. Pudenz, D. Rodríguez Pérez, A. Smull, D. Stack, M. Urbanek, R.J.M. van de Veerdonk, Z. Vendeiro, L. Wadleigh, T. Wilkason, T.Y. Wu, X. Xie, E. Zalys-Geller, X. Zhang, B.J. Bloom, PRX Quantum **5**(3), 030316 (2024). https://doi.org/10.1103/PRXQuantum.5.030316
12. H.J. Manetsch, G. Nomura, E. Bataille, K.H. Leung, X. Lv, M. Endres (2024), arXiv:2403.12021
13. G. Pichard, D. Lim, E. Bloch, J. Vaneecloo, L. Bourachot, G.J. Both, G. Mériaux, S. Dutartre, R. Hostein, J. Paris, B. Ximenez, A. Signoles, A. Browaeys, T. Lahaye, D. Dreon, Phys. Rev. Appl. **22**(2), 024073 (2024). https://doi.org/10.1103/physrevapplied.22.024073
14. S.J. Evered, D. Bluvstein, M. Kalinowski, S. Ebadi, T. Manovitz, H. Zhou, S.H. Li, A.A. Geim, T.T. Wang, N. Maskara, H. Levine, G. Semeghini, M. Greiner, V. Vuletić, M.D. Lukin, Nature (London) **622**(7982), 268 (2023). https://doi.org/10.1038/s41586-023-06481-y
15. P. Scholl, A.L. Shaw, R.B.S. Tsai, R. Finkelstein, J. Choi, M. Endres, Nature (London) **622**(7982), 273 (2023). https://doi.org/10.1038/s41586-023-06516-4, https://www.nature.com/articles/s41586-023-06516-4

16. M. Peper, Y. Li, D.Y. Knapp, M. Bileska, S. Ma, G. Liu, P. Peng, B. Zhang, S.P. Horvath, A.P. Burgers, J.D. Thompson, Phys. Rev. X. **15**(1), 011009 (2025). https://doi.org/10.1103/physrevx.15.011009, http://link.aps.org/pdf/10.1103/PhysRevX.15.011009
17. A. Cao, W.J. Eckner, T. Lukin Yelin, A.W. Young, S. Jandura, L. Yan, K. Kim, G. Pupillo, J. Ye, N. Darkwah Oppong, A.M. Kaufman, Nature (London) **634**(8033), 315 (2024). https://doi.org/10.1038/s41586-024-07913-z, https://www.nature.com/articles/s41586-024-07913-z
18. R.B.S. Tsai, X. Sun, A.L. Shaw, R. Finkelstein, M. Endres, PRX Quantum **6**(1), 010331 (2025). https://doi.org/10.1103/prxquantum.6.010331, http://link.aps.org/pdf/10.1103/PRXQuantum.6.010331
19. A.G. Radnaev, W.C. Chung, D.C. Cole, D. Mason, T.G. Ballance, M.J. Bedalov, D.A. Belknap, M.R. Berman, M. Blakely, I.L. Bloomfield, P.D. Buttler, C. Campbell, A. Chopinaud, E. Copenhaver, M.K. Dawes, S.Y. Eubanks, A.J. Friss, D.M. Garcia, J. Gilbert, M. Gillette, P. Goiporia, P. Gokhale, J. Goldwin, D. Goodwin, T.M. Graham, C.J. Guttormsson, G.T. Hickman, L. Hurtley, M. Iliev, E.B. Jones, R.A. Jones, K.W. Kuper, T.B. Lewis, M.T. Lichtman, F. Majdeteimouri, J.J. Mason, J.K. McMaster, J.A. Miles, P.T. Mitchell, J.D. Murphree, N.A. Neff-Mallon, T. Oh, V. Omole, C.P. Simon, N. Pederson, M.A. Perlin, A. Reiter, R. Rines, P. Romlow, A.M. Scott, D. Stiefvater, J.R. Tanner, A.K. Tucker, I.V. Vinogradov, M.L. Warter, M. Yeo, M. Saffman, T.W. Noel (2024), arXiv:2408.08288
20. D. Bluvstein, H. Levine, G. Semeghini, T.T. Wang, S. Ebadi, M. Kalinowski, A. Keesling, N. Maskara, H. Pichler, M. Greiner, V. Vuletić, M.D. Lukin, Nature (London) **604**(7906), 451 (2022). https://doi.org/10.1038/s41586-022-04592-6, https://www.nature.com/articles/s41586-022-04592-6
21. D. Bluvstein, S.J. Evered, A.A. Geim, S.H. Li, H. Zhou, T. Manovitz, S. Ebadi, M. Cain, M. Kalinowski, D. Hangleiter, J.P. Bonilla Ataides, N. Maskara, I. Cong, X. Gao, P. Sales Rodriguez, T. Karolyshyn, G. Semeghini, M.J. Gullans, M. Greiner, V. Vuletić, M.D. Lukin, Nature (London) **626**(7997), 58 (2024). https://doi.org/10.1038/s41586-023-06927-3
22. S. Ma, G. Liu, P. Peng, B. Zhang, S. Jandura, J. Claes, A.P. Burgers, G. Pupillo, S. Puri, J.D. Thompson, Nature (London) **622**(7982), 279 (2023). https://doi.org/10.1038/s41586-023-06438-1
23. Y. Wu, S. Kolkowitz, S. Puri, J.D. Thompson, Nat. Commun. **13**(1), 4657 (2022). https://doi.org/10.1038/s41467-022-32094-6, https://www.nature.com/articles/s41467-022-32094-6
24. I. Cong, H. Levine, A. Keesling, D. Bluvstein, S.T. Wang, M.D. Lukin, Phys. Rev. X **12**(2), 021049 (2022). https://doi.org/10.1103/PhysRevX.12.021049, https://link.aps.org/doi/10.1103/PhysRevX.12.021049
25. Q. Xu, J.P. Bonilla Ataides, C.A. Pattison, N. Raveendran, D. Bluvstein, J. Wurtz, B. Vasić, M.D. Lukin, L. Jiang, H. Zhou, Nat. Phys. **20**(7), 1084 (2024). https://doi.org/10.1038/s41567-024-02479-z, https://www.nature.com/articles/s41567-024-02479-z
26. K. Sahay, J. Jin, J. Claes, J.D. Thompson, S. Puri, Phys. Rev. X **13**(4), 041013 (2023). https://doi.org/10.1103/PhysRevX.13.041013, https://link.aps.org/doi/10.1103/PhysRevX.13.041013
27. Z. Jia, W. Huie, L. Li, W.K.C. Sun, X. Hu, Aakash, H. Kogan, A. Karve, J.Y. Lee, J.P. Covey, Npj Quantum Inf. **10**(1), 1 (2024). https://doi.org/10.1038/s41534-024-00898-7, https://www.nature.com/articles/s41534-024-00898-7
28. M. Saffman, J. Phys. B: At. Mol. Opt. Phys. **49**(20), 202001 (2016). https://doi.org/10.1088/0953-4075/49/20/202001, https://iopscience.iop.org/article/10.1088/0953-4075/49/20/202001/meta
29. E. Deist, Y.H. Lu, J. Ho, M.K. Pasha, J. Zeiher, Z. Yan, D.M. Stamper-Kurn, Phys. Rev. Lett. **129**(20), 203602 (2022). https://doi.org/10.1103/PhysRevLett.129.203602
30. T.M. Graham, L. Phuttitarn, R. Chinnarasu, Y. Song, C. Poole, K. Jooya, J. Scott, A. Scott, P. Eichler, M. Saffman, Phys. Rev. X **13**(4), 041051 (2023). https://doi.org/10.1103/PhysRevX.13.041051, https://link.aps.org/doi/10.1103/PhysRevX.13.041051
31. M.A. Norcia, W.B. Cairncross, K. Barnes, P. Battaglino, A. Brown, M.O. Brown, K. Cassella, C.A. Chen, R. Coxe, D. Crow, J. Epstein, C. Griger, A.M.W. Jones, H. Kim, J.M.

Kindem, J. King, S.S. Kondov, K. Kotru, J. Lauigan, M. Li, M. Lu, E. Megidish, J. Marjanovic, M. McDonald, T. Mittiga, J.A. Muniz, S. Narayanaswami, C. Nishiguchi, R. Notermans, T. Paule, K.A. Pawlak, L.S. Peng, A. Ryou, A. Smull, D. Stack, M. Stone, A. Sucich, M. Urbanek, R.J.M. van de Veerdonk, Z. Vendeiro, T. Wilkason, T.Y. Wu, X. Xie, X. Zhang, B.J. Bloom, Phys. Rev. X **13**(4), 041034 (2023). https://doi.org/10.1103/PhysRevX.13.041034, https://link.aps.org/doi/10.1103/PhysRevX.13.041034
32. J.W. Lis, A. Senoo, W.F. McGrew, F. Rönchen, A. Jenkins, A.M. Kaufman, Phys. Rev. X **13**(4), 041035 (2023). https://doi.org/10.1103/PhysRevX.13.041035, https://link.aps.org/doi/10.1103/PhysRevX.13.041035
33. D. Das, S. Barthwal, A. Banerjee, V. Natarajan, Phys. Rev. A **72**(3), 032506 (2005). https://doi.org/10.1103/PhysRevA.72.032506, https://link.aps.org/doi/10.1103/PhysRevA.72.032506
34. K. Pandey, A.K. Singh, P.V.K. Kumar, M.V. Suryanarayana, V. Natarajan, Phys. Rev. A **80**(2), 022518 (2009). https://doi.org/10.1103/PhysRevA.80.022518, https://link.aps.org/doi/10.1103/PhysRevA.80.022518
35. K. Singh, S. Anand, A. Pocklington, J.T. Kemp, H. Bernien, Phys. Rev. X **12**(1), 011040 (2022). https://doi.org/10.1103/PhysRevX.12.011040, https://link.aps.org/doi/10.1103/PhysRevX.12.011040
36. K. Singh, C.E. Bradley, S. Anand, V. Ramesh, R. White, H. Bernien, Science **380**(6651), 1265 (2023). https://doi.org/10.1126/science.ade5337
37. S. Anand, C.E. Bradley, R. White, V. Ramesh, K. Singh, H. Bernien, Nat. Phys. **20**(11), 1744 (2024). https://doi.org/10.1038/s41567-024-02638-2, https://www.nature.com/articles/s41567-024-02638-2
38. Y. Zeng, P. Xu, X. He, Y. Liu, M. Liu, J. Wang, D.J. Papoular, G.V. Shlyapnikov, M. Zhan, Phys. Rev. Lett. **119**(16), 160502 (2017). https://doi.org/10.1103/PhysRevLett.119.160502
39. A.G. Fowler, M. Mariantoni, J.M. Martinis, A.N. Cleland, Phys. Rev. A **86**(3), 032324 (2012). https://doi.org/10.1103/PhysRevA.86.032324, https://link.aps.org/doi/10.1103/PhysRevA.86.032324
40. A. Jenkins, J.W. Lis, A. Senoo, W.F. McGrew, A.M. Kaufman, Phys. Rev. X **12**(2), 021027 (2022). https://doi.org/10.1103/PhysRevX.12.021027, https://link.aps.org/doi/10.1103/PhysRevX.12.021027
41. S. Ma, A.P. Burgers, G. Liu, J. Wilson, B. Zhang, J.D. Thompson, Phys. Rev. X **12**(2), 021028 (2022). https://doi.org/10.1103/PhysRevX.12.021028, https://link.aps.org/doi/10.1103/PhysRevX.12.021028
42. N. Schine, A.W. Young, W.J. Eckner, M.J. Martin, A.M. Kaufman, Nat. Phys. **18**(9), 1067 (2022). https://doi.org/10.1038/s41567-022-01678-w, https://www.nature.com/articles/s41567-022-01678-w
43. R. Yamamoto, J. Kobayashi, T. Kuno, K. Kato, Y. Takahashi, New J. Phys. **18**(2), 023016 (2016). https://doi.org/10.1088/1367-2630/18/2/023016, https://iopscience.iop.org/article/10.1088/1367-2630/18/2/023016/meta
44. S. Saskin, J.T. Wilson, B. Grinkemeyer, J.D. Thompson, Phys. Rev. Lett. **122**(14), 143002 (2019). https://doi.org/10.1103/PhysRevLett.122.143002
45. T. Tomita, S. Nakajima, Y. Takasu, Y. Takahashi, Phys. Rev. A **99**(3), 031601(R) (2019). https://doi.org/10.1103/PhysRevA.99.031601, https://link.aps.org/doi/10.1103/PhysRevA.99.031601
46. D. Okuno, Y. Nakamura, T. Kusano, Y. Takasu, N. Takei, H. Konishi, Y. Takahashi, J. Phys. Soc. Jpn. **91**(8), 084301 (2022). https://doi.org/10.7566/JPSJ.91.084301
47. C. Sheng, J. Hou, X. He, K. Wang, R. Guo, J. Zhuang, B. Mamat, P. Xu, M. Liu, J. Wang, M. Zhan, Phys. Rev. Lett. **128**(8), 083202 (2022). https://doi.org/10.1103/PhysRevLett.128.083202
48. Z.J. Tao, L.G. Yu, P. Xu, J.Y. Hou, X.D. He, M.S. Zhan, Chin. Physics Lett. **39**(8), 083701 (2022). https://doi.org/10.1088/0256-307x/39/8/083701, https://iopscience.iop.org/article/10.1088/0256-307X/39/8/083701
49. W. Tian, W.J. Wee, A. Qu, B.J.M. Lim, P.R. Datla, V.P.W. Koh, H. Loh, Phys. Rev. Appl. **19**(3), 034048 (2023). https://doi.org/10.1103/physrevapplied.19.034048, https://link.aps.org/doi/10.1103/PhysRevApplied.19.034048

50. A.L. Shaw, P. Scholl, R. Finklestein, I.S. Madjarov, B. Grinkemeyer, M. Endres, Phys. Rev. Lett. **130**(19), 193402 (2023). https://doi.org/10.1103/PhysRevLett.130.193402
51. T. Grünzweig, A. Hilliard, M. McGovern, M.F. Andersen, Nat. Phys. **6**(12), 951 (2010). https://doi.org/10.1038/nphys1778, https://www.nature.com/articles/nphys1778
52. M.O. Brown, T. Thiele, C. Kiehl, T.W. Hsu, C.A. Regal, Phys. Rev. X **9**(1), 011057 (2019). https://doi.org/10.1103/PhysRevX.9.011057, https://link.aps.org/doi/10.1103/PhysRevX.9.011057
53. R.J. LeRoy, R.B. Bernstein, J. Chem. Phys. **52**(8), 3869 (1970). https://doi.org/10.1063/1.1673585, https://pubs.aip.org/aip/jcp/article-pdf/52/8/3869/18867820/3869_1_online.pdf
54. R.V. Brooks, S. Spence, A. Guttridge, A. Alampounti, A. Rakonjac, L.A. McArd, J.M. Hutson, S.L. Cornish, New J. Phys. **23**(6), 065002 (2021). https://doi.org/10.1088/1367-2630/ac0000, https://iopscience.iop.org/article/10.1088/1367-2630/ac0000/pdf
55. K.M. Jones, E. Tiesinga, P.D. Lett, P.S. Julienne, Rev. Mod. Phys. **78**(2), 483 (2006). https://doi.org/10.1103/RevModPhys.78.483, https://link.aps.org/doi/10.1103/RevModPhys.78.483
56. J. Weiner, V.S. Bagnato, S. Zilio, P.S. Julienne, Rev. Mod. Phys. **71**(1), 1 (1999). https://doi.org/10.1103/RevModPhys.71.1, https://link.aps.org/doi/10.1103/RevModPhys.71.1
57. K. Enomoto, M. Kitagawa, S. Tojo, Y. Takahashi, Phys. Rev. Lett. **100**(12), 123001 (2008). https://doi.org/10.1103/PhysRevLett.100.123001
58. J.M. Blatt, J. Comput. Phys. **1**(3), 382 (1967). https://doi.org/10.1016/0021-9991(67)90046-0, https://www.sciencedirect.com/science/article/pii/0021999167900460
59. M. Borkowski, R. Ciuryło, P.S. Julienne, R. Yamazaki, H. Hara, K. Enomoto, S. Taie, S. Sugawa, Y. Takasu, Y. Takahashi, Phys. Rev. A **84**(3), 030702(R) (2011).https://doi.org/10.1103/PhysRevA.84.030702, https://link.aps.org/doi/10.1103/PhysRevA.84.030702
60. K.K. Ni, S. Ospelkaus, M.H.G. de Miranda, A. Pe'er, B. Neyenhuis, J.J. Zirbel, S. Kotochigova, P.S. Julienne, D.S. Jin, J. Ye, Science **322**(5899), 231 (2008). https://doi.org/10.1126/science.1163861
61. A. Chotia, B. Neyenhuis, S.A. Moses, B. Yan, J.P. Covey, M. Foss-Feig, A.M. Rey, D.S. Jin, J. Ye, Phys. Rev. Lett. **108**(8), 080405 (2012). https://doi.org/10.1103/PhysRevLett.108.080405
62. M. Kitagawa, K. Enomoto, K. Kasa, Y. Takahashi, R. Ciuryło, P. Naidon, P.S. Julienne, Phys. Rev. A **77**(1), 012719 (2008). https://doi.org/10.1103/PhysRevA.77.012719, https://link.aps.org/doi/10.1103/PhysRevA.77.012719
63. Y. Yu, K. Wang, J.D. Hood, L.R.B. Picard, J.T. Zhang, W.B. Cairncross, J.M. Hutson, R. Gonzalez-Ferez, T. Rosenband, K.K. Ni, Phys. Rev. X **11**(3), 031061 (2021). https://doi.org/10.1103/PhysRevX.11.031061, https://link.aps.org/doi/10.1103/PhysRevX.11.031061
64. L.R. Liu, J.D. Hood, Y. Yu, J.T. Zhang, N.R. Hutzler, T. Rosenband, K.K. Ni, Science **360**(6391), 900 (2018). https://doi.org/10.1126/science.aar7797
65. L.R. Liu, J.D. Hood, Y. Yu, J.T. Zhang, K. Wang, Y.W. Lin, T. Rosenband, K.K. Ni, Phys. Rev. X **9**(2), 021039 (2019). https://doi.org/10.1103/PhysRevX.9.021039, https://link.aps.org/doi/10.1103/PhysRevX.9.021039
66. J.T. Zhang, Y. Yu, W.B. Cairncross, K. Wang, L.R.B. Picard, J.D. Hood, Y.W. Lin, J.M. Hutson, K.K. Ni, Phys. Rev. Lett. **124**(25), 253401 (2020). https://doi.org/10.1103/PhysRevLett.124.253401
67. L. Anderegg, L.W. Cheuk, Y. Bao, S. Burchesky, W. Ketterle, K.K. Ni, J.M. Doyle, Science **365**(6458), 1156 (2019). https://doi.org/10.1126/science.aax1265
68. X. He, K. Wang, J. Zhuang, P. Xu, X. Gao, R. Guo, C. Sheng, M. Liu, J. Wang, J. Li, G.V. Shlyapnikov, M. Zhan, Science **370**(6514), 331 (2020). https://doi.org/10.1126/science.aba7468, https://www.science.org/doi/10.1126/science.aba7468
69. D.K. Ruttley, A. Guttridge, S. Spence, R.C. Bird, C.R. Le Sueur, J.M. Hutson, S.L. Cornish, Phys. Rev. Lett. **130**(22), 223401 (2023). https://doi.org/10.1103/PhysRevLett.130.223401
70. J.W. Park, Z.Z. Yan, H. Loh, S.A. Will, M.W. Zwierlein, Science **357**(6349), 372 (2017). https://doi.org/10.1126/science.aal5066, https://www.science.org/doi/pdf/10.1126/science.aal5066

71. P.D. Gregory, J.A. Blackmore, S.L. Bromley, J.M. Hutson, S.L. Cornish, Nat. Phys. **17**(10), 1149 (2021). https://doi.org/10.1038/s41567-021-01328-7, https://www.nature.com/articles/s41567-021-01328-7
72. R. Sawant, J.A. Blackmore, P.D. Gregory, J. Mur-Petit, D. Jaksch, J. Aldegunde, J.M. Hutson, M.R. Tarbutt, S.L. Cornish, New J. Phys. **22**(1), 013027 (2020).https://doi.org/10.1088/1367-2630/ab60f4, https://iopscience.iop.org/article/10.1088/1367-2630/ab60f4/meta
73. V.V. Albert, J.P. Covey, J. Preskill, Phys. Rev. X **10**(3), 031050 (2020). https://doi.org/10.1103/PhysRevX.10.031050, https://link.aps.org/doi/10.1103/PhysRevX.10.031050
74. B.J. Furey, Z. Wu, M. Isaza-Monsalve, S. Walser, E. Mattivi, R. Nardi, P. Schindler, Quantum **8**, 1578 (2024). https://doi.org/10.22331/q-2024-12-27-1578
75. W.B. Cairncross, J.T. Zhang, L.R.B. Picard, Y. Yu, K. Wang, K.K. Ni, Phys. Rev. Lett. **126**(12), 123402 (2021). https://doi.org/10.1103/PhysRevLett.126.123402
76. G. Reinaudi, C.B. Osborn, M. McDonald, S. Kotochigova, T. Zelevinsky, Phys. Rev. Lett. **109**(11), 115303 (2012). https://doi.org/10.1103/PhysRevLett.109.115303, https://journals.aps.org/prl/pdf/10.1103/PhysRevLett.109.115303
77. S. Stellmer, B. Pasquiou, R. Grimm, F. Schreck, Phys. Rev. Lett. **109**(11), 115302 (2012). https://doi.org/10.1103/PhysRevLett.109.115302, https://journals.aps.org/prl/pdf/10.1103/PhysRevLett.109.115302
78. S. Kato, R. Yamazaki, K. Shibata, R. Yamamoto, H. Yamada, Y. Takahashi, Phys. Rev. A **86**(4), 043411 (2012). https://doi.org/10.1103/physreva.86.043411, https://journals.aps.org/pra/pdf/10.1103/PhysRevA.86.043411
79. J.D. Thompson, T.G. Tiecke, A.S. Zibrov, V. Vuletić, M.D. Lukin, Phys. Rev. Lett. **110**(13), 133001 (2013). https://doi.org/10.1103/PhysRevLett.110.133001
80. X. Zhang, K. Beloy, Y.S. Hassan, W.F. McGrew, C.C. Chen, J.L. Siegel, T. Grogan, A.D. Ludlow, Phys. Rev. Lett. **129**(11), 113202 (2022). https://doi.org/10.1103/PhysRevLett.129.113202, https://link.aps.org/doi/10.1103/PhysRevLett.129.113202
81. T. Ishiyama, K. Ono, T. Takano, A. Sunaga, Y. Takahashi, Phys. Rev. Lett. **130**(15), 153402 (2023). https://doi.org/10.1103/PhysRevLett.130.153402, https://journals.aps.org/prl/pdf/10.1103/PhysRevLett.130.153402
82. A. Kawasaki, T. Kobayashi, A. Nishiyama, T. Tanabe, M. Yasuda, Phys. Rev. A **109**(6), 062806 (2024). https://doi.org/10.1103/physreva.109.062806, https://journals.aps.org/pra/pdf/10.1103/PhysRevA.109.062806
83. S. Kato, K. Inaba, S. Sugawa, K. Shibata, R. Yamamoto, M. Yamashita, Y. Takahashi, Nat. Commun. **7**, 11341 (2016). https://doi.org/10.1038/ncomms11341
84. A.V. Taichenachev, V.I. Yudin, C.W. Oates, C.W. Hoyt, Z.W. Barber, L. Hollberg, Phys. Rev. Lett. **96**(8), 083001 (2006). https://doi.org/10.1103/PhysRevLett.96.083001
85. Z.W. Barber, C.W. Hoyt, C.W. Oates, L. Hollberg, A.V. Taichenachev, V.I. Yudin, Phys. Rev. Lett. **96**(8), 083002 (2006). https://doi.org/10.1103/PhysRevLett.96.083002
86. A. Bergschneider, V.M. Klinkhamer, J.H. Becher, R. Klemt, G. Zürn, P.M. Preiss, S. Jochim, Phys. Rev. A **97**(6), 063613 (2018). https://doi.org/10.1103/PhysRevA.97.063613, https://link.aps.org/doi/10.1103/PhysRevA.97.063613
87. J.M. Auger, S. Bergamini, D.E. Browne, Phys. Rev. A **96**(5), 052320 (2017). https://doi.org/10.1103/PhysRevA.96.052320, http://link.aps.org/pdf/10.1103/PhysRevA.96.052320
88. A. Márton, J.K. Asbóth, Quantum **7**, 1116 (2023). https://doi.org/10.22331/q-2023-09-21-1116, https://quantum-journal.org/papers/q-2023-09-21-1116/pdf/
89. B. Nikolov, E. Diamond-Hitchcock, J. Bass, N.L.R. Spong, J.D. Pritchard, Phys. Rev. Lett. **131**(3), 030602 (2023). https://doi.org/10.1103/PhysRevLett.131.030602
90. C.M. Holland, Y. Lu, L.W. Cheuk, Science **382**(6675), 1143 (2023). https://doi.org/10.1126/science.adf4272
91. D.K. Ruttley, T.R. Hepworth, A. Guttridge, S.L. Cornish, Nature **637**(8047), 827 (2025). https://doi.org/10.1038/s41586-024-08365-1
92. D.K. Ruttley, A. Guttridge, T.R. Hepworth, S.L. Cornish, PRX Quantum **5**(2), 020333 (2024). https://doi.org/10.1103/prxquantum.5.020333, http://link.aps.org/pdf/10.1103/PRXQuantum.5.020333

93. L.R.B. Picard, A.J. Park, G.E. Patenotte, S. Gebretsadkan, D. Wellnitz, A.M. Rey, K.K. Ni, Nature, 1–6 (2024). https://doi.org/10.1038/s41586-024-08177-3, https://www.nature.com/articles/s41586-024-08177-3
94. C.M. Holland, Y. Lu, S.J. Li, C.L. Welsh, L.W. Cheuk (2024), arXiv:2406.02391
95. V.A. Dzuba, V.V. Flambaum, S. Schiller, Phys. Rev. A **98**(2), 022501 (2018). https://doi.org/10.1103/PhysRevA.98.022501, https://link.aps.org/doi/10.1103/PhysRevA.98.022501
96. G. Unnikrishnan, P. Ilzhöfer, A. Scholz, C. Hölzl, A. Götzelmann, R.K. Gupta, J. Zhao, J. Krauter, S. Weber, N. Makki, H.P. Büchler, T. Pfau, F. Meinert, Phys. Rev. Lett. **132**(15), 150606 (2024). https://doi.org/10.1103/PhysRevLett.132.150606
97. S. Pucher, V. Klüsener, F. Spriestersbach, J. Geiger, A. Schindewolf, I. Bloch, S. Blatt, Phys. Rev. Lett. **132**(15), 150605 (2024). https://doi.org/10.1103/PhysRevLett.132.150605, https://journals.aps.org/prl/pdf/10.1103/PhysRevLett.132.150605
98. S. Jandura, G. Pupillo, Quantum **6**(712), 712 (2022). https://doi.org/10.22331/q-2022-05-13-712, https://quantum-journal.org/papers/q-2022-05-13-712/pdf/
99. F. Hummel, S. Weber, J. Mögerle, H. Menke, J. King, B. Bloom, S. Hofferberth, M. Li, Phys. Rev. A **110**(4), 042821 (2024). https://doi.org/10.1103/physreva.110.042821, http://link.aps.org/pdf/10.1103/PhysRevA.110.042821
100. P.M. Ireland, D.M. Walker, J.D. Pritchard, Phys. Rev. Res. **6**(1), 013293 (2024). https://doi.org/10.1103/PhysRevResearch.6.013293, https://link.aps.org/doi/10.1103/PhysRevResearch.6.013293

Chapter 7
Development of a High-Power Ultraviolet Laser System and Observation of Fast Coherent Rydberg Excitation of Ytterbium

Rydberg interactions within neutral atom arrays have been utilized in quantum simulations of diverse physical systems [1–9], as well as in the generation of entangled states in quantum information applications [10–16]. These Rydberg states–highly excited electronic states—are generally reached via a two-photon excitation in alkali atoms such as rubidium or cesium. However, this approach can lead to undesirable scattering from the intermediate state, which in turn may degrade the fidelity of Rydberg excitation [17].

In contrast, alkali-earth or alkali-earth-like atoms (AEAs) such as strontium [18–24] and ytterbium [25–29] have recently attracted considerable interest. The optical tweezer systems employing the AEA enable Rydberg excitation via a single-photon process from a metastable excited state, offering a pathway to higher-fidelity operations [30–35]. Achieving further enhancements in fidelity hinges on employing high-power lasers characterized by low frequency noise [34, 36, 37]. Nevertheless, this presents a technical obstacle: the wavelengths required for Rydberg excitation in AEAs fall within the ultraviolet (UV) spectrum, making the realization of such high-power, low-noise lasers particularly challenging.

This chapter describes the development of a high-power ultraviolet laser system and the observation of fast coherent Rydberg excitation of ytterbium, based on the published paper [38].

7.1 Development of a High-Power Ultraviolet Laser System

7.1.1 Overview of the Laser System

The transition wavelength from the metastable $^3P_2$ state to a Rydberg state in Yb is approximately 325 nm [26]. To achieve high-power output at this wavelength, we employ a two-stage frequency doubling scheme (Fig. 7.1). The laser system is

Y. Nakamura, *Generation of Highly Controllable Ytterbium Atom Array*,
Springer Theses, https://doi.org/10.1007/978-981-95-2836-3_7

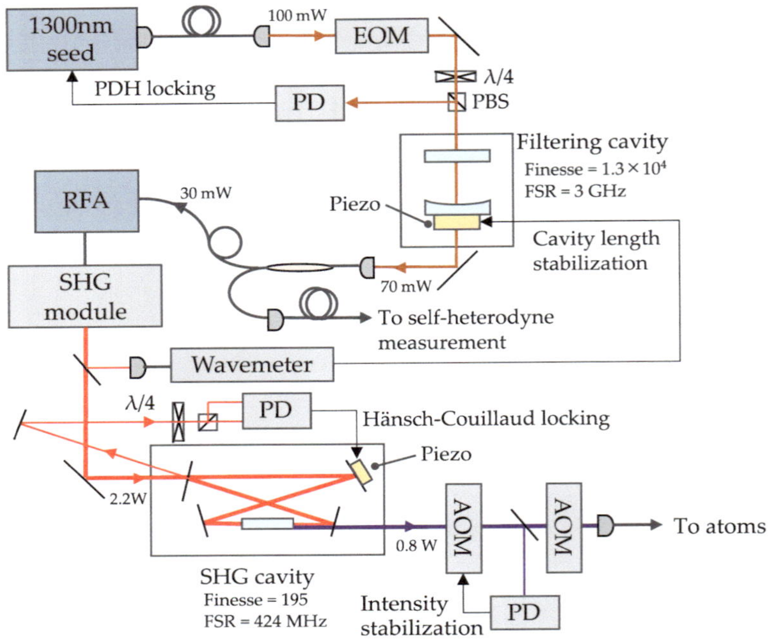

Fig. 7.1 Overview of the UV laser system. Modified from my previously published work [38]

composed of an external cavity diode laser and a tapered amplifier as a seed laser (λ = 1300 nm, Toptica, TA pro), followed by a Raman fiber amplifier coupled with an SHG module (MPB Communications, VRFA-P-3000-651.2-SF-Plus), which converts the wavelength from 1300 to 650 nm. The second stage of frequency doubling is achieved by developing an efficient SHG cavity system utilizing a beta barium borate (BBO) crystal, which further converts the wavelength from 650 to 325 nm. The seed laser, delivering 100 mW of output power, is initially coupled into a single-mode fiber. It then passes through an electro-optic modulator (EOM) before being directed into a Fabry-P'erot cavity for frequency stabilization via the Pound-Drever-Hall (PDH) locking method [39]. In addition to stabilizing the frequency, the cavity functions as a noise filter, effectively suppressing frequency fluctuations to enhance the fidelity of Rydberg excitation [14]. The cavity employed in this setup features a finesse of 1.3×10^4 and a free spectral range (FSR) of 3 GHz, yielding a 20 dB reduction in frequency noise of the transmitted light at a 1.5 MHz offset, as detailed later.

The cavity length is adjustable via piezoelectric transducers mounted between the mirror-holder tubes and is actively stabilized to the target frequency using a wavemeter (HighFinesse, WS8-10). This tunability allows the seed laser to be locked to various Rydberg resonances, thereby enabling system optimization for specific quantum protocols. The cavity exhibits a transmission efficiency of approximately 70%, which aligns well with theoretical expectations based on the mirror coatings [40]. Importantly, we observe no signs of thermal drift in the cavity, which could otherwise arise from high intra-cavity optical power.

The transmitted light from the filtering cavity, with an output power of 70 mW, is coupled into a single-mode fiber and subsequently split using a 1×2 fiber splitter with a 50:50 ratio. One arm of the splitter, delivering approximately 30 mW of power, is directed to the RFA, while the other arm is employed in a self-heterodyne measurement for frequency noise evaluation. The 1300 nm light, amplified by the RFA, is then converted to 2.5 W of 650 nm light via the SHG module.

The converted 650 nm light is sent to the SHG cavity we developed, which consists of four mirrors (three mirrors with a reflectivity $R > 99.8\%$ and an input coupler with a transmittance $T_1 = 2.1\%$ at 650 nm, and all mirrors $R < 2\%$ at 325 nm) and a BBO crystal (3 mm (height) $\times$3 mm (width) $\times$ 10 mm (length), with cut angles of $\theta = 36.7°$, $\phi = 90.0°$). The nonlinear coefficient of the BBO crystal E_{nl} is $9.4(1) \times 10^{-5}$ /W, measured by a single-pass SHG setup. The cavity finesse and FSR are 195 and 424 MHz, respectively. Figure 7.2a shows the output power of the SHG cavity under Hänsch-Couillaud locking [41]. We obtain more than 800 mW output, which is sufficient for fast Rydberg excitation at MHz-order (Fig. 7.2b). By fitting this result with the theoretical SHG output power P_{SHG} given as [42, 43]

$$P_{\mathrm{SHG}} = E_{\mathrm{nl}} P_{\mathrm{ic}}^2, \tag{7.1}$$

with

$$P_{\mathrm{ic}} = \frac{\eta T_1 P_{\mathrm{in}}}{(1 - \sqrt{(1-T_1)(1-l_{\mathrm{cav}})(1-E_{\mathrm{nl}} P_{\mathrm{ic}})})^2}, \tag{7.2}$$

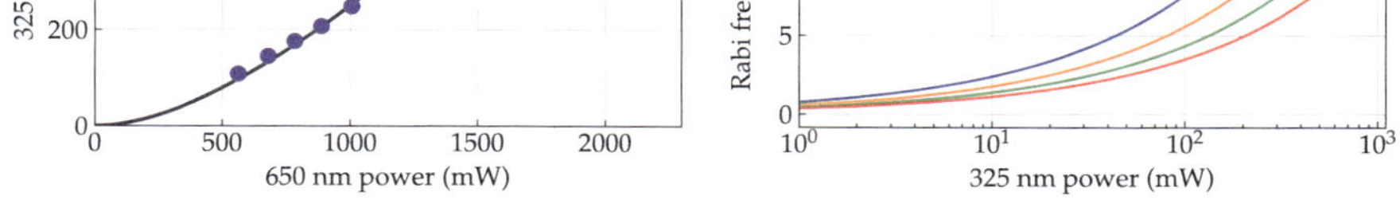

Fig. 7.2 Output power of the UV laser system. **a** The 325 nm output power after the SHG cavity we developed. The maximum power exceeds 800 mW. The dashed line is the fitting curve with Eqs. (7.1) and (7.2) with the fitting parameter of l_{cav}. With the independently measured values of $E_{\mathrm{nl}} = 9.4(1) \times 10^{-5}$ /W, $T_1 = 2.1\%$, and $\eta = 82\%$, we estimate $l_{\mathrm{cav}} = 1.08(8)\%$, consistent with the cavity finesse of 195. **b** Expected Rabi frequencies with various 325 nm beam powers. In the calculation, the reduced dipole matrix elements between $6p$ and ns states are taken from previous research [44], and a beam radius of 70 μm at the atom position is assumed. Modified from my previously published work [38]

where P_{ic} is the intracavity power of 650 nm light, l_{cav} is the round-trip power loss in the cavity except for the input coupler, η is the coupling efficiency to the cavity, and P_{in} is the input power to the cavity, we determine the cavity loss as $l_{cav} = 1.08(8)\%$, with the independently measured value of $\eta = 82\%$. The estimated l_{cav} is consistent with the cavity finesse of 195.

The output from the SHG cavity is routed through two acousto-optic modulators (AOMs): one dedicated to intensity stabilization [45], and the other for high-speed pulse shaping. The modulated light is then coupled into a single-mode fiber and delivered to the experimental chamber, where it is focused onto atoms confined in an optical tweezer array. At the atom plane, the beam has a radius of approximately 70 μm. The optical power reaching the atoms corresponds to roughly 20% of the SHG cavity output. This transmission efficiency could be improved by refining the spatial mode of the beam to achieve better overlap with the fiber mode.

7.1.2 Characteristics of the Laser Noises

Characterizing both frequency and intensity noise of the laser is essential for accurately estimating the fidelity of Rydberg excitation [34, 36]. In particular, frequency noise at or above the Rabi frequency of the Rydberg transition–typically in the MHz range–can degrade coherent excitation dynamics [14] and must therefore be carefully suppressed [34]. To address this, we employ the filtering cavity described earlier.

To quantify the suppression of frequency noise in the MHz regime, we compare delayed self-heterodyne spectra of the seed laser with and without the filtering cavity. The resulting spectra and suppression ratio are shown in Fig. 7.3. We observe more than 20 dB suppression in this frequency range, which is consistent with the cavity's linewidth (full width at half maximum) of 230 kHz.

Owing to the effective suppression of frequency noise in the megahertz range, the power spectral density (PSD) of the frequency noise at 1 MHz for the filtered seed laser is estimated to be $7 \times 10^2 \text{Hz}^2/\text{Hz}$, as determined from the phase noise analysis of the delayed self-heterodyne signals using an oscilloscope (Keysight, MXR608A) (Fig. 7.4a). Based on the analysis presented in a previous study [36], and assuming a Rabi frequency of 1 MHz, this level of noise corresponds to a Rydberg excitation error rate of approximately 8×10^{-4}–well below the threshold required for quantum error correction [46]. Further improvements could be realized by employing a seed laser with intrinsically lower frequency noise.

The relative intensity noise (RIN) spectra of 325 nm light, both with and without AOM-based intensity stabilization, are presented in Fig. 7.4b. Stabilization effectively suppresses RIN in the kHz range, which originates from the SHG cavity locking bandwidth limitations and the seed laser's frequency noise. However, the stabilization becomes less effective above 500 kHz due to the AOM's bandwidth constraints. Since the Rabi frequency exceeds 1 MHz, the Rydberg excitation error is influenced by the RIN integrated up to this range. Consequently, the intensity stabilization increases the integrated RIN up to 3 MHz to approximately 5%, slightly

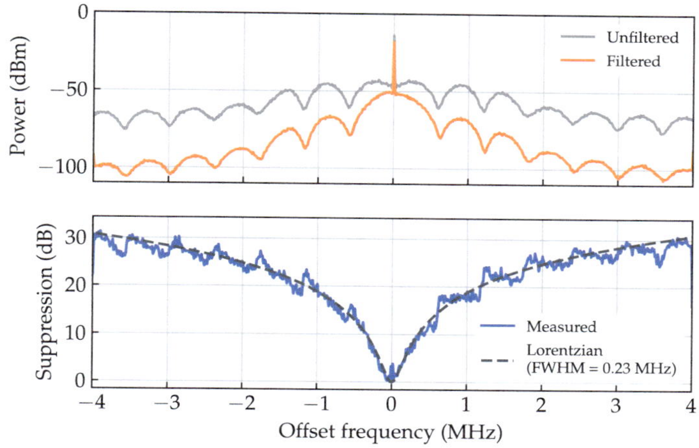

Fig. 7.3 The performance of the filtering cavity. (Top) The spectra of the delayed self-heterodyne beat notes of the seed laser with and without the filtering cavity. The periodic dip structure corresponds to the delay line length of 350 m [36]. (Bottom) The suppression factor calculated from the ratio between the spectra. As a guide for the eye, a Lorentzian function with a full width at half maximum of 230 kHz, corresponding to the filtering cavity linewidth, is plotted with a dashed line. Modified from my previously published work [38]

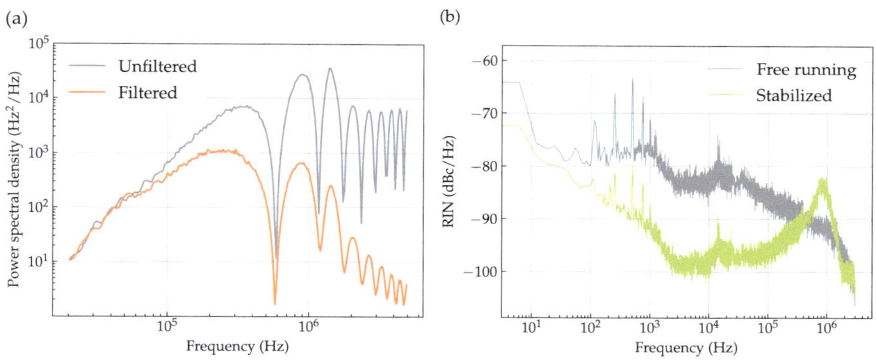

Fig. 7.4 Characteristics of the laser. **a** The frequency noise PSDs of the unfiltered and filtered seed laser, evaluated from the delayed self-heterodyne signals. The frequency noise including the servo bump around 1 MHz of the unfiltered seed laser is effectively suppressed by the filtering cavity. The measurement sensitivity decreases at multiples of the Fourier frequency equivalent to the inverse of the delay time resulting in the periodic dip structure, as well as at low Fourier frequencies. Thus, the obtained PSD underestimates the actual frequency noise in these regions. **b** The relative intensity noise spectra of 325 nm light. The intensity noise below 100 kHz is effectively suppressed by the AOM-based stabilization. The increasing noise above 500 kHz is due to the limited bandwidth of the AOM, which can be addressed by installing an EOM-based feed-forward stabilization method [47]. Modified from my previously published work [38]

higher than without stabilization. This results in a Rydberg excitation error of about 0.15%, as estimated from prior research [36]. To mitigate this, incorporating an EOM-based feed-forward stabilization technique [47] to enhance the stabilization bandwidth would be beneficial.

7.2 Coherent Excitation of Rydberg State in Ytterbium

We utilize the UV laser system to achieve coherent Rydberg excitation of individual ^{174}Yb atoms confined in an 8×8 optical tweezer array, with site spacings of 10 μm. The experimental sequence for Rydberg excitation is depicted in Fig. 7.5a. Initially, the atoms are transferred from the ground $^1S_0$ state to the $m_J = 0$ sublevel of the metastable $^3P_2$ state, as previously described. To enhance the signal-to-noise ratio in the Rydberg excitation measurements, the frequency of the 507 nm laser is chirped over a range of ±6 kHz around the $^3P_2$ resonance for 8 ms, enabling an adiabatic passage. Next, a vertically linearly polarized 325 nm beam is applied for Rydberg excitation, following the removal of residual ground-state atoms using a resonant 556 nm beam. The Rydberg atoms are then rapidly ionized by applying an electric field of 29 V/cm for 100 μs. Remaining $^3P_2$ atoms are repumped to the $^1S_0$ state using 770 and 649 nm pulses for 5 ms. Finally, the atoms in the $^1S_0$ state are imaged to determine the Rydberg excitation fraction.

We begin by conducting spectroscopy of the $(6s71s)^3S_1$ Rydberg state. For this measurement, the 325 nm beam power is set to 15 mW, and the pulse duration is 1 ms.

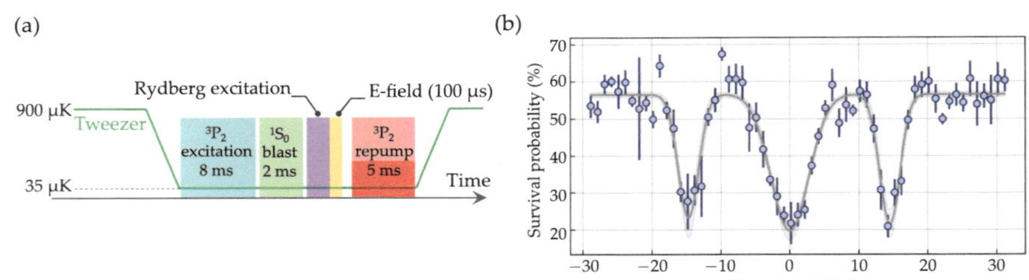

Fig. 7.5 **a** Experimental sequence. After imaging the atoms, the trapping potential is reduced to 35 μK, and a 507 nm laser with a frequency chirp of ±6 kHz around the resonance is applied for 8 ms. Any remaining atoms in the ground state are then removed by a resonant 556 nm beam. Subsequently, a 325 nm beam is used for Rydberg excitation, followed by an application of an electric field pulse of 29 V/cm. Finally, the remaining atoms in the $^3P_2$ state are repumped back to the ground state and imaged after restoring the trapping potential to its original depth. The horizontal axis is not to scale. **b** Spectrum of the $(6s71s)^3S_1$ Rydberg state. The gray line is the fitting curve with a triple-peak Gaussian function. The shaded area represents 1σ confidence intervals. The three peaks correspond to the magnetic sublevels of the Rydberg state. The error bars represent the standard error of the mean. Modified from my previously published work [38]. Modified from my previously published work [38]

Since the pulse width significantly exceeds the Rydberg state's lifetime of several tens of microseconds, the excitation fraction can be observed as atom loss even without applying an electric field. Figure 7.5b displays a typical spectrum of the Rydberg state, where the three dips correspond to the magnetic sublevels. The observed linewidth is broadened due to saturation, as the spectroscopy pulse duration is relatively long. Using the measured $^1S_0$–$^3P_2$ transition frequency of 590.902342562(63) THz [48], the transition frequency from $^1S_0$ to the $(6s71s)^3S_1$ Rydberg state is determined to be 1511.5026513(2) THz. The uncertainty in parentheses reflects the fitting error, while the overall accuracy is limited by the wavemeter's precision of 20 MHz. This result is consistent with the value reported in previous research [49].

We then proceed to coherently excite the $m_J = 0$ sublevel of the $(6s71s)^3S_1$ Rydberg state. To achieve this, the 325 nm beam power is increased to 70 mW, and the magnetic field is aligned with the beam polarization during excitation to maximize the Rabi frequency per excitation beam intensity. Figure 7.6 illustrates the observed Rabi oscillation between the $^3P_2$ and Rydberg states, obtained by varying the pulse duration from 0 to 2 μs. The measured Rabi frequency is $2\pi \times 2.13(3)$ MHz, comparable to previous demonstrations of high-fidelity two-qubit gates in neutral atom tweezer arrays [35]. Further improvements in beam shaping and alignment are expected to nearly double the Rabi frequency.

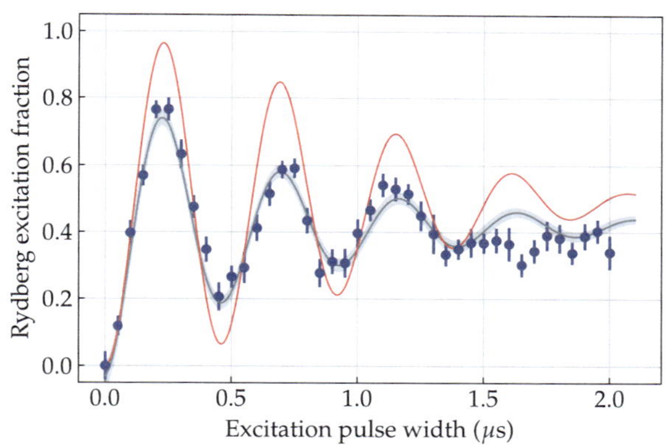

Fig. 7.6 Coherent Rabi oscillation signal. The gray line represents the fit using a damped sine function, yielding a Rabi frequency of 2.13(3) MHz. The error bars represent the standard error of the mean. The red line shows a simulated dynamics model that includes measured shot-to-shot fluctuations in both intensity and frequency. The intensity fluctuation is characterized by $\sigma_1 = 8.7\%$, and the frequency fluctuation by $\sigma_2 = 0.3$ MHz. In the simulation, the relative intensity and detuning are sampled from normal distributions $\mathcal{N}(1, \sigma_1^2)$ and $\mathcal{N}(0, \sigma_2^2)$, respectively, where $\mathcal{N}(\mu, \sigma^2)$ is a normal distribution with a mean of μ and a standard deviation of σ. The averaged dynamics are computed over 10,000 iterations. Modified from my previously published work [38]

7.3 Discussion

The observed decay in the Rabi oscillation of the Rydberg excitation is primarily attributed to shot-to-shot fluctuations in both the intensity and frequency of the excitation laser. Intensity fluctuations, measured to be approximately 9% around the mean value, likely arise from variations in the pulse shape. This occurs because the RF frequency is applied to the AOM only during UV laser irradiation to protect the fiber from high-power UV light. Consequently, the AOM does not fully thermalize before pulse generation, leading to shot-to-shot variations in the pulse area. To address this, a sample-and-hold technique [14, 30] combined with EOM-based intensity stabilization [47] could be implemented.

Frequency fluctuations, estimated to be at least 0.3 MHz, persist despite the use of a filtering cavity for laser stabilization. This estimation is based on the standard deviation of 0.3 MHz in the wavemeter readings used for cavity stabilization. The limitation stems from the wavemeter's restricted frequency resolution. This issue could be resolved by locking the seed laser to a stable frequency reference, such as an optical frequency comb stabilized by an ultranarrow laser [50].

We simulate the system dynamics while incorporating these fluctuations, with the results shown as the red line in Fig. 7.6. Although the simulated excitation fraction does not perfectly align with the experimental data–suggesting additional decoherence sources such as atom temperature, differential light shifts between the $^3P_2$ and Rydberg states, or electric field noise–the decay of the oscillation qualitatively matches the observed experimental behavior.

7.4 Conclusion

In conclusion, we have successfully developed a high-power UV laser system tailored for Rydberg excitation from the $^3P_2$ state of Yb atoms. Utilizing a two-stage frequency doubling approach with a Raman fiber amplifier and a custom-built SHG cavity, we achieved an output power exceeding 800 mW at 325 nm. To address frequency noise in the MHz range, which can degrade Rydberg excitation with MHz-order Rabi frequencies, we implemented a filtering cavity with a linewidth of 230 kHz. This cavity achieved a 20 dB suppression of the seed laser's frequency noise in the MHz range, enabling high-fidelity Rydberg excitation. Additionally, we characterized the intensity noise at 325 nm and demonstrated that AOM-based intensity stabilization effectively suppresses noise up to 500 kHz. Future improvements will focus on extending the stabilization bandwidth using EOM-based techniques [47].

Using the developed laser system, we performed spectroscopy and coherent excitation of the $(6s71s)^3S_1$ Rydberg state. The measured resonance frequency was consistent with previous studies [49]. By increasing the laser power, we observed Rabi oscillations between the $m_{J=0}$ sublevels of the $^3P_2$ and Rydberg states, with a frequency of $2\pi \times 2.13(3)$ MHz. The observed decay in the Rabi oscillations was attributed to shot-to-shot fluctuations in laser frequency and intensity, which can be mitigated using established experimental techniques.

This high-power UV laser system with effective frequency noise suppression represents a significant step toward achieving fast and high-fidelity Rydberg excitation. With further enhancements in laser noise suppression, this system holds great promise for advancing quantum simulation and computation in neutral atom arrays.

References

1. H. Labuhn, D. Barredo, S. Ravets, S. de Léséleuc, T. Macrì, T. Lahaye, A. Browaeys, Nature **534**(7609), 667 (2016). https://doi.org/10.1038/nature18274
2. H. Bernien, S. Schwartz, A. Keesling, H. Levine, A. Omran, H. Pichler, S. Choi, A.S. Zibrov, M. Endres, M. Greiner, V. Vuletić, M.D. Lukin, Nature **551**(7682), 579 (2017). https://doi.org/10.1038/nature24622
3. V. Lienhard, S. de Léséleuc, D. Barredo, T. Lahaye, A. Browaeys, M. Schuler, L.P. Henry, A.M. Läuchli, Phys. Rev. X **8**(2), 021070 (2018), https://link.aps.org/doi/10.1103/PhysRevX.8.021070
4. S. de Léséleuc, V. Lienhard, P. Scholl, D. Barredo, S. Weber, N. Lang, H.P. Büchler, T. Lahaye, A. Browaeys, Science **365**(6455), 775 (2019). https://doi.org/10.1126/science.aav9105, https://www.science.org/doi/10.1126/science.aav9105?url_ver=Z39.88-2003&rfr_id=ori:rid:crossref.org&rfr_dat=cr_pub%3dpubmed
5. G. Semeghini, H. Levine, A. Keesling, S. Ebadi, T.T. Wang, D. Bluvstein, R. Verresen, H. Pichler, M. Kalinowski, R. Samajdar, A. Omran, S. Sachdev, A. Vishwanath, M. Greiner, V. Vuletić, M.D. Lukin, Science **374**(6572), 1242 (2021). https://doi.org/10.1126/science.abi8794
6. P. Scholl, H.J. Williams, G. Bornet, F. Wallner, D. Barredo, L. Henriet, A. Signoles, C. Hainaut, T. Franz, S. Geier, A. Tebben, A. Salzinger, G. Zürn, T. Lahaye, M. Weidemüller, A. Browaeys, PRX Quantum **3**(2), 020303 (2022). https://doi.org/10.1103/PRXQuantum.3.020303, https://link.aps.org/doi/10.1103/PRXQuantum.3.020303
7. P. Scholl, M. Schuler, H.J. Williams, A.A. Eberharter, D. Barredo, K.N. Schymik, V. Lienhard, L.P. Henry, T.C. Lang, T. Lahaye, A.M. Läuchli, A. Browaeys, Nature **595**(7866), 233 (2021). https://doi.org/10.1038/s41586-021-03585-1
8. S. Ebadi, T.T. Wang, H. Levine, A. Keesling, G. Semeghini, A. Omran, D. Bluvstein, R. Samajdar, H. Pichler, W.W. Ho, S. Choi, S. Sachdev, M. Greiner, V. Vuletić, M.D. Lukin, Nature (London) **595**(7866), 227 (2021). https://doi.org/10.1038/s41586-021-03582-4
9. D. González-Cuadra, T.V. Zache, J. Carrasco, B. Kraus, P. Zoller, Phys. Rev. Lett. **129**(16), 160501 (2022). https://doi.org/10.1103/PhysRevLett.129.160501
10. D. Jaksch, J.I. Cirac, P. Zoller, S.L. Rolston, R. Cote, M.D. Lukin, Phys. Rev. Lett. **85**(10), 2208 (2000). https://doi.org/10.1103/PhysRevLett.85.2208
11. M.D. Lukin, M. Fleischhauer, R. Cote, L.M. Duan, D. Jaksch, J.I. Cirac, P. Zoller, Phys. Rev. Lett. **87**(3), 037901 (2001). https://doi.org/10.1103/PhysRevLett.87.037901, http://link.aps.org/pdf/10.1103/PhysRevLett.87.037901
12. S. Jandura, G. Pupillo, Quantum **6**(712), 712 (2022). https://doi.org/10.22331/q-2022-05-13-712, https://quantum-journal.org/papers/q-2022-05-13-712/pdf/
13. L. Isenhower, E. Urban, X.L. Zhang, A.T. Gill, T. Henage, T.A. Johnson, T.G. Walker, M. Saffman, Phys. Rev. Lett. **104**(1), 010503 (2010). https://doi.org/10.1103/PhysRevLett.104.010503, http://link.aps.org/pdf/10.1103/PhysRevLett.104.010503
14. H. Levine, A. Keesling, A. Omran, H. Bernien, S. Schwartz, A.S. Zibrov, M. Endres, M. Greiner, V. Vuletić, M.D. Lukin, Phys. Rev. Lett. **121**(12), 123603 (2018). https://doi.org/10.1103/PhysRevLett.121.123603

15. H. Levine, A. Keesling, G. Semeghini, A. Omran, T.T. Wang, S. Ebadi, H. Bernien, M. Greiner, V. Vuletić, H. Pichler, M.D. Lukin, Phys. Rev. Lett. **123**(17), 170503 (2019). https://doi.org/10.1103/PhysRevLett.123.170503
16. S.J. Evered, D. Bluvstein, M. Kalinowski, S. Ebadi, T. Manovitz, H. Zhou, S.H. Li, A.A. Geim, T.T. Wang, N. Maskara, H. Levine, G. Semeghini, M. Greiner, V. Vuletić, M.D. Lukin, Nature (London) **622**(7982), 268 (2023). https://doi.org/10.1038/s41586-023-06481-y
17. M. Saffman, J. Phys. B: At. Mol. Opt. Phys. **49**(20), 202001 (2016). https://doi.org/10.1088/0953-4075/49/20/202001, https://iopscience.iop.org/article/10.1088/0953-4075/49/20/202001/meta
18. A. Cooper, J.P. Covey, I.S. Madjarov, S.G. Porsev, M.S. Safronova, M. Endres, Phys. Rev. X **8**(4), 041055 (2018). https://link.aps.org/doi/10.1103/PhysRevX.8.041055
19. M.A. Norcia, A.W. Young, A.M. Kaufman, Phys. Rev. X **8**(4), 041054 (2018). https://doi.org/10.1103/PhysRevX.8.041054, https://link.aps.org/doi/10.1103/PhysRevX.8.041054
20. K. Barnes, P. Battaglino, B.J. Bloom, K. Cassella, R. Coxe, N. Crisosto, J.P. King, S.S. Kondov, K. Kotru, S.C. Larsen, J. Lauigan, B.J. Lester, M. McDonald, E. Megidish, S. Narayanaswami, C. Nishiguchi, R. Notermans, L.S. Peng, A. Ryou, T.Y. Wu, M. Yarwood, Nat. Commun. **13**(1), 2779 (2022). https://doi.org/10.1038/s41467-022-29977-z, https://www.nature.com/articles/s41467-022-29977-z
21. A. Urech, I.H.A. Knottnerus, R.J.C. Spreeuw, F. Schreck, Phys. Rev. Res. **4**(2), 023245 (2022). https://link.aps.org/doi/10.1103/PhysRevResearch.4.023245
22. R. Tao, M. Ammenwerth, F. Gyger, I. Bloch, J. Zeiher, Phys. Rev. Lett. **133**(1), 013401 (2024). https://doi.org/10.1103/PhysRevLett.133.013401
23. G. Unnikrishnan, P. Ilzhöfer, A. Scholz, C. Hölzl, A. Götzelmann, R.K. Gupta, J. Zhao, J. Krauter, S. Weber, N. Makki, H.P. Büchler, T. Pfau, F. Meinert, Phys. Rev. Lett. **132**(15), 150606 (2024). https://doi.org/10.1103/PhysRevLett.132.150606
24. X. Yan, C. He, K. Wen, Z. Ren, P.T.F. Wong, E. Hajiyev, G.B. Jo, Phys. Rev. Res. **7**(1), 013031 (2025). https://doi.org/10.1103/PhysRevResearch.7.013031
25. S. Saskin, J.T. Wilson, B. Grinkemeyer, J.D. Thompson, Phys. Rev. Lett. **122**(14), 143002 (2019). https://doi.org/10.1103/PhysRevLett.122.143002
26. D. Okuno, Y. Nakamura, T. Kusano, Y. Takasu, N. Takei, H. Konishi, Y. Takahashi, J. Phys. Soc. Jpn. **91**(8), 084301 (2022). https://doi.org/10.7566/JPSJ.91.084301
27. A. Jenkins, J.W. Lis, A. Senoo, W.F. McGrew, A.M. Kaufman, Phys. Rev. X **12**(2), 021027 (2022). https://doi.org/10.1103/PhysRevX.12.021027, https://link.aps.org/doi/10.1103/PhysRevX.12.021027
28. W. Huie, L. Li, N. Chen, X. Hu, Z. Jia, W.K.C. Sun, J.P. Covey, PRX Quantum **4**(3), 030337 (2023). https://doi.org/10.1103/PRXQuantum.4.030337, https://link.aps.org/doi/10.1103/PRXQuantum.4.030337
29. M.A. Norcia, W.B. Cairncross, K. Barnes, P. Battaglino, A. Brown, M.O. Brown, K. Cassella, C.A. Chen, R. Coxe, D. Crow, J. Epstein, C. Griger, A.M.W. Jones, H. Kim, J.M. Kindem, J. King, S.S. Kondov, K. Kotru, J. Lauigan, M. Li, M. Lu, E. Megidish, J. Marjanovic, M. McDonald, T. Mittiga, J.A. Muniz, S. Narayanaswami, C. Nishiguchi, R. Notermans, T. Paule, K.A. Pawlak, L.S. Peng, A. Ryou, A. Smull, D. Stack, M. Stone, A. Sucich, M. Urbanek, R.J.M. van de Veerdonk, Z. Vendeiro, T. Wilkason, T.Y. Wu, X. Xie, X. Zhang, B.J. Bloom, Phys. Rev. X **13**(4), 041034 (2023). https://doi.org/10.1103/PhysRevX.13.041034, https://link.aps.org/doi/10.1103/PhysRevX.13.041034
30. I.S. Madjarov, J.P. Covey, A.L. Shaw, J. Choi, A. Kale, A. Cooper, H. Pichler, V. Schkolnik, J.R. Williams, M. Endres, Nat. Phys. **16**(8), 857 (2020).https://doi.org/10.1038/s41567-020-0903-z, https://www.nature.com/articles/s41567-020-0903-z
31. N. Schine, A.W. Young, W.J. Eckner, M.J. Martin, A.M. Kaufman, Nat. Phys. **18**(9), 1067 (2022). https://doi.org/10.1038/s41567-022-01678-w, https://www.nature.com/articles/s41567-022-01678-w
32. P. Scholl, A.L. Shaw, R.B.S. Tsai, R. Finkelstein, J. Choi, M. Endres, Nature (London) **622**(7982), 273 (2023), https://www.nature.com/articles/s41586-023-06516-4

33. A. Cao, W.J. Eckner, T. Lukin Yelin, A.W. Young, S. Jandura, L. Yan, K. Kim, G. Pupillo, J. Ye, N. Darkwah Oppong, A.M. Kaufman, Nature **634**(8033), 315 (2024). https://doi.org/10.1038/s41586-024-07913-z, https://www.nature.com/articles/s41586-024-07913-z
34. R.B.S. Tsai, X. Sun, A.L. Shaw, R. Finkelstein, M. Endres, PRX quantum **6**(1), 010331 (2025), http://link.aps.org/pdf/10.1103/PRXQuantum.6.010331
35. M. Peper, Y. Li, D.Y. Knapp, M. Bileska, S. Ma, G. Liu, P. Peng, B. Zhang, S.P. Horvath, A.P. Burgers, J.D. Thompson, Phys. Rev. X. **15**(1), 011009 (2025).https://doi.org/10.1103/physrevx.15.011009, http://link.aps.org/pdf/10.1103/PhysRevX.15.011009
36. X. Jiang, J. Scott, M. Friesen, M. Saffman, Phys. Rev. A **107**(4), 042611 (2023). https://doi.org/10.1103/PhysRevA.107.042611, https://link.aps.org/doi/10.1103/PhysRevA.107.042611
37. S. de Léséleuc, D. Barredo, V. Lienhard, A. Browaeys, T. Lahaye, Phys. Rev. A **97**(5), 053803 (2018). https://doi.org/10.1103/PhysRevA.97.053803, https://link.aps.org/doi/10.1103/PhysRevA.97.053803
38. Y. Nakamura, N. Ozawa, T. Kusano, R. Yokoyama, K. Shibata, T. Takano, Y. Takasu, Y. Takahashi, J. Phys. Soc. Jpn. **94**, 014301 (2025). https://doi.org/10.7566/jpsj.94.014301, https://journals.jps.jp/doi/abs/10.7566/JPSJ.94.014301
39. R.W.P. Drever, J.L. Hall, F.V. Kowalski, J. Hough, G.M. Ford, A.J. Munley, H. Ward, Appl. Phys. B **31**(2), 97 (1983). https://doi.org/10.1007/bf00702605, https://link.springer.com/article/10.1007/BF00702605
40. C.J. Hood, H.J. Kimble, J. Ye, Phys. Rev. A **64**(3), 033804 (2001). https://doi.org/10.1103/PhysRevA.64.033804, https://journals.aps.org/pra/pdf/10.1103/PhysRevA.64.033804
41. T.W. Hansch, B. Couillaud, Opt. Commun. **35**(3), 441 (1980). https://doi.org/10.1016/0030-4018(80)90069-3
42. E.S. Polzik, H.J. Kimble, Opt. Lett. **16**(18), 1400 (1991). https://doi.org/10.1364/ol.16.001400, https://opg.optica.org/abstract.cfm?uri=ol-16-18-1400
43. M. Pizzocaro, D. Calonico, P.C. Pastor, J. Catani, G.A. Costanzo, F. Levi, L. Lorini, Appl. Opt. **53**(16), 3388 (2014). https://doi.org/10.1364/AO.53.003388, https://opg.optica.org/abstract.cfm?uri=ao-53-16-3388
44. J.P. Covey, A. Sipahigil, M. Saffman, Phys. Rev. A **100**(1), 012307 (2019). https://doi.org/10.1103/physreva.100.012307, http://link.aps.org/pdf/10.1103/PhysRevA.100.012307
45. F. Tricot, D.H. Phung, M. Lours, S. Guérandel, E. de Clercq, Rev. Sci. Instrum. **89**(11), 113112 (2018). https://doi.org/10.1063/1.5046852
46. A.G. Fowler, A.M. Stephens, P. Groszkowski, Phys. Rev. A **80**(5), 052312 (2009). https://doi.org/10.1103/PhysRevA.80.052312, https://journals.aps.org/pra/pdf/10.1103/PhysRevA.80.052312
47. Y. Wang, K. Wang, E.F. Fenton, Y.W. Lin, K.K. Ni, J.D. Hood, Opt. Express **28**(21), 31209 (2020). https://doi.org/10.1364/OE.405002
48. A. Yamaguchi, Metastable state of ultracold and quantum degenerate ytterbium atoms: high-resolution spectroscopy and cold collisions. Ph.D. Thesis, Kyoto University, 2008, http://yagura.scphys.kyoto-u.ac.jp/publications/thesis/PhD_yamaguchi.pdf
49. J.T. Wilson, S. Saskin, Y. Meng, S. Ma, R. Dilip, A.P. Burgers, J.D. Thompson, Phys. Rev. Lett. **128**(3), 033201 (2022). https://doi.org/10.1103/PhysRevLett.128.033201
50. H. Inaba, K. Hosaka, M. Yasuda, Y. Nakajima, K. Iwakuni, D. Akamatsu, S. Okubo, T. Kohno, A. Onae, F.L. Hong, Opt. Express **21**(7), 7891 (2013), https://opg.optica.org/abstract.cfm?uri=oe-21-7-7891

Chapter 8
Conclusion and Outlook

8.1 Conclusion

This thesis has presented the construction of a hybrid atom tweezer array of nuclear spin and optical clock qubits based on dual-isotope ytterbium atoms, as well as the development of a high-power UV laser system for Rydberg excitation towards realizing quantum error correction. We briefly review the achievements of this thesis.

- *Coherent manipulations of nuclear spin and optical clock qubits in ytterbium atoms* (Chap. 5).
  We have demonstrated efficient initialization and coherent control of nuclear spin qubits in a ^{171}Yb atom array. The nuclear spin qubits exhibit second-order long coherence times and megahertz-order fast qubit controllability, highlighting their potential as promising data qubits for quantum computing.
  Furthermore, we performed narrow-line spectroscopy and coherent manipulations of optical clock qubits in a ^{174}Yb atom array. Our results showed that the 532 nm tweezer satisfies the near-magic condition for the optical clock qubits, enabling high-fidelity state manipulations. Although the Rabi oscillations and the Ramsey signal show the limited coherence times of the optical clock qubits, these findings provide a solid foundation for the future utilization of optical clock qubits as ancilla qubits.
- *Hybrid atom tweezer array of nuclear spin and optical clock qubits* (Chap. 6) [1]
  We successfully constructed a hybrid atom tweezer array composed of ^{171}Yb and ^{174}Yb atoms. This system demonstrated atom rearrangement within the dual-Yb system and tunable loading probabilities for each isotope, highlighting its potential for molecule tweezer experiments, as well as neutral atom-based quantum computing. Our investigation revealed that the optical clock qubit readout has a negligible impact on the coherence of nuclear spin qubits, representing a critical milestone for the implementation of mid-circuit measurements in quantum error correction protocols. Furthermore, we observed minimal coherence degradation

Y. Nakamura, *Generation of Highly Controllable Ytterbium Atom Array*,
Springer Theses, https://doi.org/10.1007/978-981-95-2836-3_8

from 556 nm imaging of the optical clock qubits, suggesting that further crosstalk suppression can be achieved through optimized 556 nm imaging techniques.

- *Development of a high-power ultraviolet laser system and observation of fast coherent Rydberg excitation of ytterbium* (Chap. 7) [2].
  We developed a high-power UV laser system tailored for Rydberg excitation of ytterbium atoms. The system comprises a 1300 nm seed laser, a Raman fiber amplifier, and two frequency doubling systems. We achieved a 325 nm laser with a maximum output power of 0.8 W, enabling rapid Rydberg excitation. By incorporating a filtering cavity, the laser phase noise was suppressed by 20 dB at an offset frequency of 1.5 MHz, which is crucial for achieving high-fidelity Rydberg excitation. Using this system, we observed fast coherent Rydberg excitation of ytterbium atoms with a Rabi frequency of 2.13(3) MHz, establishing a key step towards high-fidelity two-qubit gate operations.

8.2 Outlook

Below, we outline several future directions for the hybrid atom tweezer array experiments.

- **Utilization of the $^3P_0$ state for the optical clock qubit**
  The $^3P_0$ state in ^{174}Yb atoms is a metastable excited state and serves as a promising candidate for the optical clock qubit. Owing to the absence of electronic total angular momentum J, this state is immune to vector and tensor light shifts. By employing a magic wavelength laser for the optical tweezer array, the $^3P_0$ state is expected to exhibit extended coherence times [3, 4] and enable high-fidelity state manipulations [5, 6].
- **759 nm optical tweezer array**
  This wavelength serves as a magic wavelength for the $^3P_0$ state [7–9], thereby enhancing the coherence and state manipulation fidelity of the $^3P_0$ state. Additionally, by tuning the angle between the tweezer polarization and the quantization axis, it satisfies a magic condition for the $^3P_1$ state in both ^{171}Yb [5] and ^{174}Yb [10] atoms, enabling efficient cooling and high fidelity imaging of single atoms in the tweezer array.
- **Exploring the inter-isotope Rydberg interactions and implementation of two-qubit gates**
  Although the intra-isotope Rydberg interactions of ^{171}Yb and ^{174}Yb have been extensively studied and utilized for two-qubit gates, the inter-isotope Rydberg interactions remain unexplored. By investigating these interactions with the support of theoretical calculations informed by detailed Rydberg spectroscopy research of ytterbium [11], we aim to realize two-qubit gates between nuclear spin and optical clock qubits, which are essential for quantum error correction protocols. Additionally, the potential existence of a Förster resonance in the inter-isotope Rydberg interactions could enhance the interaction range, offering a novel platform

for quantum simulation. This would leverage the imbalance between inter- and intra-isotope Rydberg interactions in a dual-isotope atom array, enabling unique opportunities for exploring complex many-body quantum systems [12–15].

- **Single heteronuclear molecule creation in the tweezer array**
  The tunability of dual occupancy of ^{171}Yb and ^{174}Yb atoms in the tweezer array offers a unique platform for the creation of single heteronuclear molecules. By employing techniques such as stimulated Raman adiabatic passage or two-photon photoassociation, we aim to produce single ground-state heteronuclear molecules within the tweezer array.
  This development opens new avenues for encoding quantum information in the rich internal states of molecules, providing access to a diverse set of quantum states for computation [16]. Furthermore, it paves the way for novel quantum error correction protocols utilizing molecular qubits, leveraging their inherent rich internal structure [17–19].
- **Implementation of quantum error correction protocols**
  Towards the ultimate goal of fault-tolerant quantum computing, we aim to implement quantum error correction protocols using the hybrid atom tweezer array. The demonstrated low-crosstalk readout of optical clock qubits, while preserving the coherence of nuclear spin qubits, positions us well to realize mid-circuit measurements essential for quantum error correction.
  By integrating the high-power UV laser system for high-fidelity Rydberg excitation and leveraging insights from inter-isotope Rydberg interaction studies, we aim to achieve high-fidelity two-qubit gates between data and ancilla qubits. These advancements will enable robust error syndrome measurements and serve as a critical step towards the realization of fault-tolerant quantum computing.

References

1. Y. Nakamura, T. Kusano, R. Yokoyama, K. Saito, K. Higashi, N. Ozawa, T. Takano, Y. Takasu, Y. Takahashi, Phys. Rev. X **14**(4), 041062 (2024). https://doi.org/10.1103/PhysRevX.14.041062, https://journals.aps.org/prx/abstract/10.1103/PhysRevX.14.041062
2. Y. Nakamura, N. Ozawa, T. Kusano, R. Yokoyama, K. Shibata, T. Takano, Y. Takasu, Y. Takahashi, J. Phys. Soc. Jpn. **94**, 014301 (2025). https://doi.org/10.7566/jpsj.94.014301, https://journals.jps.jp/doi/abs/10.7566/JPSJ.94.014301
3. M.A. Norcia, A.W. Young, W.J. Eckner, E. Oelker, J. Ye, A.M. Kaufman, Science **366**(6461), 93 (2019). https://doi.org/10.1126/science.aay0644
4. A.W. Young, W.J. Eckner, W.R. Milner, D. Kedar, M.A. Norcia, E. Oelker, N. Schine, J. Ye, A.M. Kaufman, Nature **588**(7838), 408 (2020). https://doi.org/10.1038/s41586-020-3009-y
5. J.W. Lis, A. Senoo, W.F. McGrew, F. Rönchen, A. Jenkins, A.M. Kaufman, Phys. Rev. X **13**(4), 041035 (2023). https://doi.org/10.1103/PhysRevX.13.041035, https://link.aps.org/doi/10.1103/PhysRevX.13.041035
6. J.A. Muniz, M. Stone, D.T. Stack, M. Jaffe, J.M. Kindem, L. Wadleigh, E. Zalys-Geller, X. Zhang, C.A. Chen, M.A. Norcia, J. Epstein, E. Halperin, F. Hummel, T. Wilkason, M. Li, K. Barnes, P. Battaglino, T.C. Bohdanowicz, G. Booth, A. Brown, M.O. Brown, W.B. Cairncross, K. Cassella, R. Coxe, D. Crow, M. Feldkamp, C. Griger, A. Heinz, A.M.W. Jones, H. Kim, J. King, K. Kotru, J. Lauigan, J. Marjanovic, E. Megidish, M. Meredith, M. McDonald, R. Morshead, S. Narayanaswami, C. Nishiguchi, T. Paule, K.A. Pawlak, K.L. Pudenz, D.R.

Pérez, A. Ryou, J. Simon, A. Smull, M. Urbanek, R.J.M. van de Veerdonk, Z. Vendeiro, T.Y. Wu, X. Xie, B.J. Bloom (2024), arXiv:2411.11708
7. Z.W. Barber, C.W. Hoyt, C.W. Oates, L. Hollberg, A.V. Taichenachev, V.I. Yudin, Phys. Rev. Lett. **96**(8), 083002 (2006). https://doi.org/10.1103/PhysRevLett.96.083002
8. N.D. Lemke, A.D. Ludlow, Z.W. Barber, T.M. Fortier, S.A. Diddams, Y. Jiang, S.R. Jefferts, T.P. Heavner, T.E. Parker, C.W. Oates, Phys. Rev. Lett. **103**(6), 063001 (2009). https://doi.org/10.1103/PhysRevLett.103.063001, https://journals.aps.org/prl/pdf/10.1103/PhysRevLett.103.063001
9. V.A. Dzuba, A. Derevianko, J. Phys. B At. Mol. Opt. Phys. **43**(7), 074011 (2010). https://doi.org/10.1088/0953-4075/43/7/074011, https://iopscience.iop.org/article/10.1088/0953-4075/43/7/074011/meta
10. T.O. Höhn, R.A. Villela, E. Zu, L. Bezzo, R.M. Kroeze, M. Aidelsburger (2024), arXiv:2412.14163
11. M. Peper, Y. Li, D.Y. Knapp, M. Bileska, S. Ma, G. Liu, P. Peng, B. Zhang, S.P. Horvath, A.P. Burgers, J.D. Thompson, Phys. Rev. X. **15**(1), 011009 (2025). https://doi.org/10.1103/physrevx.15.011009, http://link.aps.org/pdf/10.1103/PhysRevX.15.011009
12. L. Homeier, A. Bohrdt, S. Linsel, E. Demler, J.C. Halimeh, F. Grusdt, Commun. Phys. **6**(1), 1 (2023). https://doi.org/10.1038/s42005-023-01237-6, https://www.nature.com/articles/s42005-023-01237-6
13. N. Chepiga, Phys. Rev. Lett. **132**(7), 076505 (2024). https://doi.org/10.1103/PhysRevLett.132.076505, https://journals.aps.org/prl/pdf/10.1103/PhysRevLett.132.076505
14. C. Li, S. Liu, H. Wang, W. Zhang, Z.X. Li, H. Zhai, Y. Gu, Phys. Rev. Lett. **133**(22), 223401 (2024). https://doi.org/10.1103/PhysRevLett.133.223401
15. A.H. Nevidomskyy, H. Bernien, A. Canright (2024), arXiv:2407.05885
16. R. Sawant, J.A. Blackmore, P.D. Gregory, J. Mur-Petit, D. Jaksch, J. Aldegunde, J.M. Hutson, M.R. Tarbutt, S.L. Cornish, New J. Phys. **22**(1), 013027 (2020). https://doi.org/10.1088/1367-2630/ab60f4, https://iopscience.iop.org/article/10.1088/1367-2630/ab60f4/meta
17. V.V. Albert, J.P. Covey, J. Preskill, Phys. Rev. X **10**(3), 031050 (2020). https://doi.org/10.1103/PhysRevX.10.031050, https://link.aps.org/doi/10.1103/PhysRevX.10.031050
18. S.P. Jain, E.R. Hudson, W.C. Campbell, V.V. Albert, Phys. Rev. Lett. **133**(26), 260601 (2024). https://doi.org/10.1103/PhysRevLett.133.260601, https://journals.aps.org/prl/pdf/10.1103/PhysRevLett.133.260601
19. B.J. Furey, Z. Wu, M. Isaza-Monsalve, S. Walser, E. Mattivi, R. Nardi, P. Schindler, Quantum **8**, 1578 (2024). https://doi.org/10.22331/q-2024-12-27-1578

MIX
Papier aus verantwortungsvollen Quellen
Paper from responsible sources
FSC® C105338

If you have any concerns about our products,
you can contact us on
ProductSafety@springernature.com

In case Publisher is established outside the EU,
the EU authorized representative is:
Springer Nature Customer Service Center GmbH
Europaplatz 3, 69115 Heidelberg, Germany

Printed by Libri Plureos GmbH
in Hamburg, Germany